The Time of Revolution

Bloomsbury Studies in Continental Philosophy

Bloomsbury Studies in Continental Philosophy is a major monograph series from Bloomsbury Academic. The series features first-class scholarly research monographs across the field of Continental philosophy. Each work makes a major contribution to the field of philosophical research.

Adorno's Concept of Life, Alastair Morgan
Adorno's Poetics of Critique, Steven Helmling
Badiou and Derrida, Antonio Calcagno
Badiou, Marion and St Paul, Adam Miller
Being and Number in Heidegger's Thought, Michael Roubach
Crisis in Continental Philosophy, Robert Piercey
Deleuze and Guattari, Fadi Abou-Rihan
Deleuze and Guattari's Philosophy of History, Jay Lampert
Deleuze and the Genesis of Representation, Joe Hughes
Derrida, Simon Morgan Wortham
Derrida and Disinterest, Sean Gaston
Derrida: Ethics Under Erasure, Nicole Anderson
Domestication of Derrida, Lorenzo Fabbri
Encountering Derrida, Simon Morgan Wortham
Foucault's Heidegger, Timothy Rayner
Foucault's Legacy, C.G. Prado
Gabriel Marcel's Ethics of Hope, Jill Graper Hernandez
Gadamer and the Question of the Divine, Walter Lammi
Gilles Deleuze, Constantin V. Boundas
Heidegger and a Metaphysics of Feeling, Sharin N. Elkholy
Heidegger and Authenticity, Mahon O'Brien
Heidegger and Happiness, Matthew King
Heidegger and Philosophical Atheology, Peter S. Dillard
Heidegger and the Place of Ethics, Michael Lewis
Heidegger Beyond Deconstruction, Michael Lewis
Heidegger, Politics and Climate Change, Ruth Irwin
Heidegger's Early Philosophy, James Luchte
In the Shadow of Phenomenology, Stephen H. Watson
Irony of Heidegger, Andrew Haas
Kant, Deleuze and Architectonics, Edward Willatt
Merleau-Ponty's Phenomenology, Kirk M. Besmer
Michel Henry, Jeffrey Hanson

Nietzsche and the Anglo-Saxon Tradition, Louise Mabille
Nietzsche's Ethical Theory, Craig Dove
Nietzsche's Thus Spoke Zarathustra, James Luchte
Phenomenology, Institution and History, Stephen H. Watson
Ricoeur and Lacan, Karl Simms
Sartre's Phenomenology, David Reisman
Simultaneity and Delay, Jay Lampert
Thinking Between Deleuze and Kant, Edward Willatt
Who's Afraid of Deleuze and Guattari?, Gregg Lambert
Zizek and Heidegger, Thomas Brockelman

The Time of Revolution

Kairos and Chronos in Heidegger

Felix Ó Murchadha

B L O O M S B U R Y
LONDON • NEW DELHI • NEW YORK • SYDNEY

Bloomsbury Academic
An imprint of Bloomsbury Publishing Plc

50 Bedford Square
London
WC1B 3DP
UK

175 Fifth Avenue
New York
NY 10010
USA

www.bloomsbury.com

First published 2013

British Library Cataloguing-in-Publication Data
A catalogue record for this book is available from the British Library.

ISBN: HB: 978-1-4411-0246-1

Library of Congress Cataloging-in-Publication Data
Ó Murchadha, Felix.
The time of revolution : kairos and chronos in Heidegger / Felix Ó Murchadha.
p. cm.
Includes bibliographical references (p.   ) and index.
ISBN 978-1-4411-0246-1 (hardcover : alk. paper)– ISBN 978-1-4411-2016-8 (epub)– ISBN 978-1-4411-1945-2 (ebook) 1. Heidegger, Martin, 1889-1976. Sein und Zeit. 2. Ontology. 3. Space and time. I. Title.
B3279.H48S46654513 2012
193–dc23
2012020413

Typeset by Fakenham Prepress Solutions, Fakenham, Norfolk NR21 8NN
Printed and bound in Great Britain

To my wife Anne and my son Felix Alexander

Contents

Acknowledgements

This is a translation and a reworking of a book published in 1999 in German, entitled *Zeit des Handelns und Möglichkeit der Verwandlung. Kairologie und Chronologie bei Heidegger im Jahrzehnt nach 'Sein und Zeit'* (*Time of Action and the Possibility of Transformation. Kairology and Chronology in Heidegger during the decade following 'Being and Time'*). The author is grateful to the publishers of that book, Königshausen und Neumann Verlag (Würzburg), for permission to publish this revised translation. The change in title reflects a shift of emphasis, in which the notion of revolution is given more prominence as a key concept in Heidegger's thought. Each chapter has been revised in content and style in terms of this change of emphasis. Account has also been taken of publications since 1999, both of new volumes in the Heidegger *Gesamtausgabe* and of secondary literature on Heidegger, especially that published in English. In addition, Chapter 1 (on kairos and chronos) includes a new section on the transcendental ambitions of *Being and Time*, and Chapter 5 (on the 'work of politics') contains an extended engagement with the debate regarding the significance of Heidegger's political involvement with the National Socialists. Some of the material from Chapter 2 has been published in *Philosophy Today*.

The first version of the present book was submitted as my doctoral dissertation at the *Bergische Universität Wuppertal* (Germany). I am forever grateful to my *Doctorvater*, Prof. Klaus Held, for his guidance and encouragement, not only during my doctoral studies, but ever since. Anyone familiar with his thought will see his influence throughout this book. In Wuppertal and Bochum universities, I benefited from discussions with my fellow students and lecturers. Prof. Heinrich Hüni, Prof. László Tengelyi and Prof. Burkhard Liebsch greatly helped me to develop and deepen my understanding of Heidegger.

Prof. Markus Wörner (National University of Ireland, Galway) first sparked my interest in the problem of time and the question of the 'kairos'; for this, and for his support and encouragement over many years, I owe him a debt of gratitude.

Prof. Will McNeill's work has been a constant source of inspiration, and his comments on sections of the present book were invaluable for me.

I am also grateful to the anonymous reviewers at Continuum for their most helpful advice.

Over the years I have been encouraged from time to time by colleagues and graduate students to publish this book in a revised translation in English. I am grateful for their encouragement.

For their help in the final preparation of this book, I would like to thank especially two friends who have seen this project through from its very inception many years ago and who generously helped me in the preparation of this book, Marty Fairbairn and Anthony Jenkins.

Finally, for their patience at my many absences while working on this book and their love, support and help, I am very grateful to my wife and son, Anne and Felix Alexander to whom this book is dedicated.

Abbreviations

I have used the following abbreviations for *Being and Time* and for the German originals of Heidegger's works.

BT *Being and Time: a Translation of Sein und Zeit.* Trans. by J. Stambaugh. Albany: SUNY Press, 2006

EiM *Einführung in die Metaphysik.* Tübingen: Niemeyer, 1987

GA 19 *Platons Sophistes. Gesamtausgabe 19*, I. Schüßler (ed.). Frankfurt a.M.: Klostermann, 1992

GA 20 *Prolegomena zur Geschichte des Zeitbegriffs. Gesamtausgabe 20*, P. Jaeger (ed.). Frankfurt a. M: Klosermann, 1988

GA 21 *Logic: die Frage nach der Wahrheit. Gesamtausgabe 21*, W. Biemel (ed.). Frankfurt a.M.: Klostermann, 1995

GA 24 *Die Grundprobleme der Phänomenologie. Gesamtausgabe 24* (second edition), F-W. von Hermann (ed.). Frankfurt a. M: Klosermann, 1989

GA 26 *Die metaphysiche Anfangsgründe der Logic. Gesamtausgabe 26*, K. Held (ed.). Frankfurt a.M.: Klostermann,1978

GA 29/30 *Die Grundbegriffe der Metaphysik. Welt – Endlichkeit – Einsamkeit. Gesamtausgabe 29/30*, F-W. von Herrmann (ed.). Frankfurt a.M.: Klostermann, 1992

GA 31 *Vom Wesen der menschlichen Freiheit. Gesamtausgabe 31*, H. Tietjen (ed.). Frankfurt a.M.: Klostermann, 1982

GA 34 *Vom Wesen der Wahrheit. Gesamtausgabe 34*, H. Mörchen (ed.). Frankfurt a.M.: Klostermann, 1988

GA 38 *Logik als Frage nach des Wesens der Sprache. Gesamtausgabe 38*, G. Seubold (ed.). Frankfurt a.M.: Klostermann, 1998

GA 45 *Grundfragen der Philosophie. Ausgewählte „Probleme" der „Logik".*

Gesamtausgabe 45 (second edition), F-W von Herrmann (ed.). Frankfurt a.M.: Klostermann, 1992

GA 60 *Phänomenologie des religiösen Lebens. Gesamtausgabe 60*, M. Jung, T. Regehly and C. Strube (eds). Frankfurt a.M.: Klostermann, 1995

GA 64 'Der Begriff der Zeit', in *Der Begriff der Zeit. Gesamtausgabe 64*, F.W. von Hermann (ed.). Frankfurt a.M.: Klostermann, 2004, pp. 105–25.

GA 65 *Beiträge zur Philosophie. Von Ereignis. Gesamtausgabe 65*, F-W. von Hermann (ed.). Frankfurt a.M.: Klostrermann, 1989.

H *Sein und Zeit* (seventh edition). Tübingen: Niemeyer, 1986

Hw *Holzwege*. Frankfurt a.M.: Klostermann, 1980

KPM *Heidegger und das Problem der Metaphysik*. Frankfurt a. M.: Klostermann, 1991

KV 'Kasseler Vorträge'. Bröcker and Rodi (eds). *Dilthey Jahrbuch*, 8 (1992/3), 143-180.

Logica *Logica: Leciones de M. Heidegger (semester verano 1934) en el legado de Helene Weiss*. Bilingual edition, V. Farias (ed.). Barcelona: Anthropos. 1991

R *Das Rektorat 1933/34. Tatsachen und Gedanken*. Frankfurt a.M.: Klostermann, 1983

US *Unterwegs zur Sprache* Stuttgart: Klett-Cotta, 2007

Wm *Wegmarken* (second edition). Frankfurt a.M.: Klostermann, 1978

ZS *Zollikoner Seminare Protokolle, Zwiegespräche, Briefe*, M. Boss (ed.). Frankfurt a.M.: Klostermann, 1994

ZSD Zu Sache des Denkens. Tübingen: Niemeyer.

Introduction

> *No great historical event is better calculated than the French Revolution to teach political writers and statesmen to be cautious in their speculations; for never was any such event, stemming for factors so far back in the past, so inevitable and yet so completely unforeseen.*
>
> Tocqueville, 1971, p. 33

Revolutions in general, not only historical instances such as the French Revolution, call practical reason into question. Human beings normally live and act in accordance with the self-evident assumption that the future will mark a continuation of the order with which they are familiar. However, human action itself opens up the possibility of the new; it can open up a possible future, which will be other than the past. We do not know *what* this future will bring. Similarly *when* such new possibilities will open up is unknown. We know little concerning the *contents* of this future. However, we can not only know about this future, we must have such knowledge. This is because the time of such novel emergence is a decisive, critical time – a kairological time – in which a new order becomes possible, in which new possibilities for life, knowledge and the whole of human conduct open up, but it is also a time in which new misfortunes become possible. Such critical time is of philosophical interest, because it is a unique time that does not allow itself be subsumed under timeless categories. The challenge that a revolution poses concerns not only political thinkers, but also philosophers more generally. This is so because, if it is the case that a kairologically new time is really unforeseeable, then it cannot be ordered within a chronological context by a philosophy of history. As a unique event, the revolution is a temporal phenomenon which cannot be either explained or justified in terms of a chronological order.

In this context, revolution is not just to be understood in political terms as a reversal of existing political institutions, but rather as meaning a transformation of the world. The difficulty of understanding the essence of revolution are hinted at in the quotation from de Tocqueville's *The Ancient Régime and The French Revolution,* which stands above as the epigraph for this introduction. In retrospect, that which is experienced in the revolutionary moment as fully surprising and sudden, appears as somehow inevitable. What occurs between this before and after? Do we know more concerning the conditions of the event, which perhaps developed through long periods in the past, or is it rather the case that a revolution is an occurrence which brings about a transformation in its very appearance of novelty, such that which was previously unforeseeable appears as inevitable? If the latter is the case, this would be a very remarkable situation, and this book will attempt to understand how this might be possible through an analysis of Heidegger's discussion of temporality and historicity.

Martin Heidegger was a revolutionary thinker. He not only thought revolution, but did so in a revolutionary manner, and he notoriously attempted – for ten months in 1933–4 as Rector of Freiburg University – to contribute to bringing a revolution into effect. This political engagement was without doubt a blunder. Two principle errors lay at its basis. Firstly, Heidegger was the victim of the reverse side of de Toqueville's observation: revolutions occur suddenly and surprisingly, but by the same token may not occur despite every expectation. Secondly, he did not recognize the modal character of revolutions, and in this he went against the tendency of his own work: a revolution is a possibility which can never be 'actualized'. Revolution remains always possibility. This does not mean that revolutions do not happen. They certainly do, but not as actuality, rather as possibility. Revolutions do not take effect, are not actualized, they make possible. It is Heidegger's own transformation of the concept of possibility in *Being and Time* which can make sense of this.

Nonetheless, the present book is not primarily aimed at contributing to the debate concerning Heidegger's political engagement with the National Socialists. Heidegger's actions in the years 1933–4 will be discussed in Chapter 5, but only with a view to understand and critically discuss the concept of historical time which informed these actions. The relevance of the latter to his politics was already suggested by Heidegger himself in 1936 – as reported by his former student Karl Löwith (cf. Löwith, 1993, p. 142). The debate regarding Heidegger's politics will be addressed with a view to understanding his politics philosophically rather than – as with certain commentators, above all Emmanuel Faye – understanding his philosophy politically. This is done neither to defend nor to exonerate Heidegger: those categories, which unfortunately frame much of the debate on this issue, belong in the courtroom not in philosophy. The goal is rather to understand Heidegger's political engagement in terms of his account of historical time, and specifically of revolutionary, kairological time.

Historicity is a *temporal* phenomenon. It is a principle thesis of this work that Heidegger understood time not from the experience of measurement and of measures – as, for example, Aristotle did – but rather from the experience of acting. This means that the 'original time', the 'initiating (*anfängliche*)' time – time in its own emerging – is traceable in action. In saying this, little seems to be said, since measuring is also an action. Measuring is, though, a theoretical operation. Every theoretical operation refers to an object before it, which is an object in its presence. Time, however, as Heidegger understands it, is – to speak roughly and provisionally for now – that which surpasses presence: time, so to speak, sways between being and non-being. For the theoretical approach – as Augustine has already shown in book 11 of the *Confessions* – any understanding of time suffers from fundamental aporias. On the other hand, action in the sense of doing (*praxis*) and making/producing (*poiesis*) relates necessarily to time; namely, to the past and the future. In doing and making, humans experience a world which changes from the past into the future. The temporal phenomena of emerging and passing away are fundamental to doing and making.

A methodological consequence flows from this, namely that time can only be studied in the context of these phenomena; it cannot be isolated as a theoretical problem, but rather must be understood in the context of human action, its goals and its boundaries.

The present book is an attempt – to use Heidegger's own formulation – 'to understand time from time'. (Heidegger, 2007a, p. 200; GA 64, p. 107) This formulation can of course seem misleading: the question concerns time, but time is already presumed in the question. Thus, we appear to be moving in a circle. This circle can only be justified if the 'time' which we are asking about and the time from which we are asking are different. Heidegger's main thesis is that there is such a split in time and that time must be understood in terms of this split, not for instance in terms of the timeless. The present book puts forward the thesis that the split in question here is one between chronological and kairological time. In this case, the fundamental question is how these two forms of time relate to one another.

'Chronos' and 'kairos' are two Greek words for that which is called 'time' in English, '*Zeit*' in German and '*temps*' in French.[1] 'Kairos' is a qualitative concept of time, which means the opportune point in time: the opportune time to do something, the right time to act. 'Chronos', on the other hand, is for the most part a quantitative concept of time. It is not by accident that 'chronology' or 'chronicle' means an account of what happened during a particular period of time. As this suggests, 'chronos' is not to be understood purely quantitatively. In the context of human action, we experience chronos as continuity and kairos as a moment of vision – *Augenblick* – that breaks with the continuity, as an other time, as a time which is opportune for action in the emphatic sense. In kairos, the discontinuity of time appears.

I will discuss these concepts of time in more detail in Chapter 1; for now, I wish only to outline the following investigation. This book concerns the relation between chronos and kairos. In dealing with this question, we encounter certain problems which every investigation of time is forced to deal with. In the first place, the question concerning time does not lend itself to a direct approach. (cf. Theunissen, 1991, p. 37) Time is not an object which can be brought before the investigator and then examined. It is precisely in our familiarity with it that the mystery of time lies for us.[2] There is no direct way from the time in which we live to the time we are seeking to understand. Nevertheless, every experience is temporally constituted. In every experience there is a transition, a movement, which encloses future, past and present. For the most part, but precisely not always, we experience this transition as a continuity.

Sometimes this continuity is interrupted. In such a case the transition itself becomes questionable. The transition takes place in an interplay between – to use the terms employed by Reinhart Koselleck – 'the space of experience (*Erfahrungsraum*) and the horizon of experience (*Erfahrungshorizont*)'. (cf. Koselleck, 1985, pp. 267–88) 'Behind' us lies the space of our experience; 'before' us lies the horizon of our expectations. Expectations arise out of experience. In this way, the horizon of our expectations is bound by our experiences. On the other hand, the space of our experiences is being continually ordered anew in terms of our expectations: in rough terms, we can say that we see the positive elements of our experiences if we have hope for the future, and only the negative elements if we are in despair. But when we are dealing with a new event, something which contradicts our past experiences, something which opens up a new horizon and which refigures retrospectively our space of experience, then – in that moment – the *world* will be suddenly otherwise. It is not accidental that, following such events, the need arises to narrate the past and to give an account of it, because that which was normal has become suddenly an anomaly: the previously self-evident 'chronology' must now, in a new world, become newly chronologically ordered, newly 'chronologized'.

Such chronologizing is considered by many contemporary philosophers and historians to be the fundamental character of human time. (cf. Carr, 1986; White, 1987; and above all Ricoeur, 1984) Paul Ricoeur stated this position clearly and it formed the major thesis of his three volume work, *Time and Narrative*. As he puts it: '*time becomes human to the extent to which it is articulated through a narrative mode, and narrative attains its full meaning when it becomes a condition of temporal existence.*' (Ricoeur, 1984, p. 52 [emphasis in original]) 'Narrated time' is human time, and this means that human beings experience time from and through poiesis. Practice (in Ricoeur's schema mimesis I) is only the fore-structure of the narrated configuration (mimesis II), and receives its fully human temporal structure only though narration by means of refiguration (mimesis III). Indeed, Ricoeur's use of mimesis and mimetic structures clearly show the poietical character of time in his account.

This poietical experience of time always rests on the chronological. The poietical experience is based on a temporal order, which must always be produced again (through an always new narration), and in this way retains an inner continuity. However, in the experiences of rupture and of the new, this continuity is called into question, indeed it is torn apart. The experience of rupture is the experience of the irruption of *dis*continuity in the existing order. This experience will be termed kairological in the present work. This kairological form of experience will be interpreted on the basis of the practical (as opposed to poietical) experience of time.

The intention of the present investigation is not, however, simply to develop an opposing position to the poietical understanding of time, as if we could merely oppose praxis to poiesis. Rather, the aim of this work is to investigate the close relationship of time to human action in the dynamic interrelation of praxis and poiesis. It is indeed the case that in the poietical mode of discussing time, the kairos is largely ignored, or at the very least subordinated to chronos. The philosophical question concerning the kairos is, in part, motivated by the fact that the kairological experience is a practical experience which is obscured by poietical interpretations of time. As such, the questions concerning kairos and praxis are closely related. However, in the course of this investigation, it will become clear that poiesis has kairological characteristics, and also that praxis is structured chronologically. The 'split' in time between 'kairos' and 'chronos' involves equally praxis and poiesis.

Neither form of temporality can be explained without reference to the other. In the case of kairos, this reference is clear: a *dis*continuity is a dis*continuity*. Chronos on the other hand, does not refer immediately to kairos; it appears at first that continuity can be understood without any reference to its possible fragmentation. It would appear that only kairos is referred to chronos and not vice versa; kairos appears as a mere epiphenomenon. However, against this the phenomenological evidence is that the experience of time is never purely chronological. The threat of discontinuity always emerges anew within the chronological experience itself. A continuity without such a threat of rupture would no longer be a continuity, would not be chronology, because the chronological is itself the continual overcoming of temporal difference: without the threat of that difference exploding into rupture there would never be any need for poietical action (for example, the narrating of stories).

It is thus the case that the inner connection and relationship of kairos and chronos cannot be based on any priority of chronos. Rather, in the present work, the relation of these temporal forms will be understood – speaking metaphorically – as a crossing in the sense of a crossing-over of two ways, the way of chronos and the way of kairos. This crossing over will be understood as a dynamic interrelation where, in a unique 'situation', the possibilities of continuity and discontinuity are encountered. In a 'situation', in this emphatic sense of the word, there occurs – and indeed 'in a moment' – a transformation that cannot be overtaken or subsumed by any existing chronology. This is so because it cannot be explained in terms of causes or effects. Such a transformation is always only subsequent; that is, in terms of a *new* chronology, integrated into a renewed continuity.

I have already referred to time as ordered and ordering. Without time there would be no order, because only in the motion of the coming to itself and the passing away does multiplicity happen – only in becoming is there a dispersal into the many[3] – and the possibility of order only arises as a 'coming together of this and that according to a rule' (Waldenfels, 1987, p. 17); that is, in a multiplicity. The time which is opened up by the possibility of dispersal is also the possibility of order. The possibility of order, as the possibility, which goes beyond the still monism of the present, is time as both multiple and unifying. This characteristic of time brings Husserl to the conclusion that time itself does not move – 'Time is fixed (*starr*) and yet time flows' (Husserl, 1964, p. 67); long before that, it lay at the basis of Augustine's understanding of time as an interplay of *distentio* and *intentio*. (Augustine, *Confessions* XI, 17–39; cf. Flask, 1993, pp. 338–402)

Augustine represents a classical view of the phenomenon of time. It is necessary, however, in going beyond the classical position to stress that, while time is ordering, it also has within it the possibility of the chaotic. The split between chronos and kairos brings with it an interplay of order and chaos. This tendency to chaos has indeed, at least since Augustine, been noted in philosophical reflections on time. While time is only in the unity of past, present and future, there still is between these modes of time a difference out of which a dispersal of that unity can result. However, this chaotic moment has repeatedly been obscured through the emphasis on chronology. If time is understood as chronos, then the chaotic moment can be disguised, because a continual everyday order of time is assumed.

The question of the interplay of chronos and kairos relates closely to the relationship of temporality to historicity. Historicity names the possibility of transformation *and* continuity. The human being is historical because in her unique actions she can allow events to occur, which change her and her world. At the same time, she has the *task*, precisely because such events occur, to bring about continuity with the past. Through the experience of unique events, the human being finds her past to be foreign and strange. As L.P. Hartley famously expressed it, '[T]he past is a foreign country, they do things differently there'. (Hartley, 2002, p. 17) This strangeness does not necessarily make all continuity impossible. On the contrary, it can open up previously unknown and unexplored possibilities in the past, which in turn can form a new continuity.

It is clear that the temporal structure of kairos and chronos is operative at the level of historicity. For this reason, the attempt to found historicity on temporality fails. In *Being and Time*, Heidegger made such an attempt. In the present work, it will be shown that temporality – both as kairological and as chronological – is already contained within historicity. This means firstly, that the difference between kairos and chronos can only be recognized through historical – unique – events, and secondly, that chronology as well as kairology is constituted intersubjectively. Hence, in both chronos and kairos the singular experience of the individual is based in a communal – and as such historical – experience of time.

The question of time is addressed in this work in relation to Heidegger's thought. He, more than any other philosopher in the past century, allows us to rethink the traditional account of time. The interpretation of Heidegger's texts brings its own problems, which are not simply a matter of their inherent difficulty, but that they force us to confront a thought, which endangers the very project of such an interpretative endeavour from the beginning. Heidegger himself cast doubt on the sense of any such attempt. In a letter to Otto Pöggeler (dated 17 April 1964), he expressed the hope that the time would come when one would no longer write *about* Heidegger, but rather engage in a substantial discussion concerning *die Sache* (the matters themselves).[4] Nevertheless, in justifying my own project, I can call on Heidegger's own mode of philosophizing. When he is engaged in interpreting Aristotle, Kant, Hegel, etc., the difference between interpretation and substantial discussion is no longer apparent. To express this in terms of a typically Heideggerian distinction, he sought not to give a correct (*richtig*) interpretation, but rather to draw from the text its truth (*Wahrheit*). The truth is that which the text tells us concerning *die Sache*.

One could, however, make the following objection: while Aristotle, Kant and Hegel were philosophers and their texts as such of a philosophical nature, Heidegger is not a philosopher in the same sense. According to his own self-interpretation, his thinking goes beyond philosophy, and understands itself as indicating the path of further thought. As such, according to such an objection, it is inappropriate to interpret him in the same manner as traditional philosophers. Such an objection rests on a false assumption: Heidegger's thought does not consist in 'doctrines' as one might find, say, in Aristotle, but Heidegger himself did not interpret philosophers in the tradition in relation to their doctrines. Rather, what he sought was to make the unsaid and the unthought in a philosopher's work explicit. Heidegger's interpretative approach was not to consider the answers a philosopher gave, but rather their hidden questions, and that which gets lost in the philosopher's own striving for answers. In other words, Heidegger was not interested in the finished and fixed doctrine, but rather in such possibilities for thinking which were hidden in a philosophical text. It must therefore be legitimate to apply the same interpretive principle to Heidegger himself. We are not true to Heidegger as a philosopher if we attempt to interpret him in an orthodox 'Heideggerian' manner; the philosophical task is much more concerned with bringing to light that which is to be thought in his texts.

The aim of these investigations is *not* to interpret Heidegger 'correctly'. To interpret Heidegger is to go beyond his texts in order to understand the phenomena (*die Sache*) which his texts bring to light. Methodologically, this means interpreting his texts not as objects of consideration, but as guides on the path of thinking. The aim of this work is not in the first instance to criticize, supplement or, for that matter, to interpret Heidegger, although in the course of these investigations he will be criticized, supplemented and interpreted. The aim is rather, through a repetition of his texts, to pose the question concerning the relation of kairos and chronos. Repetition (*Wieder-holung*) is understood here in Heidegger's sense (cf. Heidegger, 1990, pp. 163–6; KPM, pp. 238–42) – which itself is influenced by Nietzsche and Kierkegaard – as the revitalizing re- (*wieder*) trieval (*holen*) of an event. In this case, the event is Heidegger's own texts, which in their uniqueness cannot be subsumed under a generalizing thought, but can only be worked through again anew.

In this working through, some ways will be taken with Heidegger. At the beginning stands his 'first magnum opus' *Being and Time* (1927). The path will be taken to the threshold of what Pöggeler calls his 'second magnum opus': *Contributions to Philosophy* (1936–8). This period of roughly a decade begins with the breakthrough of *Being and Time*, but also contains its failure, goes through the crisis of his thinking in 1929–30, sees the beginning of the turning (*Kehre*), witnesses his turn to art and to Hölderlin, displays the beginnings of his readings of Nietzsche, and in its midst covers his political engagement of 1933–4. This period in his thinking has an experimental character; there is a constant attempt to find new ways. He finds that new way in the *Contributions*. In the present book, Heidegger's renewed attempts to pursue that way from the late 1930s through to the 1950s will not be discussed. The aim is rather to show the ways in which Heidegger attempted to go beyond *Being and Time*, specifically in terms of his rethinking of the relation of praxis and poiesis, kairos and chronos.

In the course of these investigations, the term 'the other beginning' (*anderer Anfang*), which first appears in the *Contributions*, will be used. This term is not taken from that work and so to speak imposed from the outside on the earlier texts; rather, it arises organically from a consideration of the latter works. The 'other beginning' is a phenomenon of revolution; it refers to another time, a time to begin, and a time which begins, a time which is alien to us. It is a time of the stranger, the alien, the foreigner (*Fremde*), in which we ourselves become strange and foreign to our previous selves.

Before giving a brief outline of the chapters, I need to address some more reservations concerning the project undertaken here. This book follows more or less the chronological order of Heidegger's texts from 1927–36. Reiner Schürmann has argued that one can only understand Heidegger's earlier work if one reads them in reverse chronological order – that is, read the early work in the light of the later. (Schürmann, 1990, pp. 13–14) One of his reasons for pursuing this course is the observation that, if one reads Heidegger's early work without reference to his later, his concept of praxis can appear as totalitarian, while if one takes the opposite course the anarchic character of the early concept of praxis becomes apparent. It is, however, not clear that a chronological reading does lead to a totalitarian account of praxis, as the present work will attempt to show. Furthermore, to read Heidegger from the later to early work runs the risk of reading him too much in terms of his own self-interpretation, which must always remain *one* interpretative possibility amongst others.

A further objection is that to begin with *Being and Time* today, after the publication of Heidegger's early Freiburg and Marburg lectures, is an arbitrary decision: *Being and Time* needs to be read now in the light of these early lectures, and any consideration of Heidegger should begin there. In response to this, it must first be pointed out that account is taken in this book of the early lecture courses in interpreting *Being and Time*. Nevertheless, it is true that no extended interpretation of these lectures is attempted, and that they are read from the point of view of *Being and Time*. This is justified because *Being and Time* remains undoubtedly the main result of his philosophical reflections in this period, and it is of all his works the one to which he returned most until the end of his life. For these reasons, this book makes the attempt to understand temporality and historicity starting from *Being and Time*.

Being and Time failed in its attempt, it remains a fragment. How we answer the question as to why it remains a fragment depends on how we interpret *Being and Time* itself and how we interpret the 'turning' in Heidegger's thinking from the 1930s on. In the following work, the failure will be understood in the context of the discussion of temporality and historicity.

Chapter 1 will show that the attempt to found historicity on temporality of Dasein – an attempt which lies at the basis of the project of fundamental ontology – fails. The reason for this, it is argued, is that Heidegger's account of authentic time is not clearly understood in terms of the relation of chronological and kairological time. If 'authentic time' was understood kairologically, then it would have to be conceived as historical.

The latter thesis is developed in Chapter 2 through a discussion of the difference between praxis and poiesis. This chapter shows the relation between the analysis of time and the concept of action in *Being and Time*. Heidegger's account of the ready-to-hand is interpreted as a 'destructuring (*Destruktion*)' reading of Aristotle's account of poiesis. The analyses of mood, mortality and the priority of the future are given a close reading. It is argued that the emphasis on the singular, ownmost existence is one-sided and that, kairologically understood, time is fundamentally intersubjective and communal.

To understand the time of revolution, it is necessary to think of freedom in such a way that it is not conceived on the basis of causality. Kairological time emerges suddenly, without cause (understood in chronological terms). The ancient problem of the origins of time and the temporal nature of the causal relation underlying this origin can be understood differently, when one thinks of it kairologically and on the basis of action, rather than cosmologically. It is not accidental that in the years following *Being and Time*, the question of freedom comes to enjoy a central place in Heidegger's thought. Examining key texts from 1929 to 1936, it is shown that not only does Heidegger give us the means to think of freedom as the basis of the emerging of time, particularly kairological time, but that he shows the connection between this question of freedom at the basis of time and that of truth: with the emerging of time in a kairological moment, a new order of entities is disclosed. As such, one should speak not only of a 'revolutionary time,' but also of 'revolutionary truth.'

Chapter 4 returns to the question of the interplay of praxis and poiesis. In the years after *Being and Time*, poiesis played an increasingly important role. The 'work' (*Werk*) which in *Being and Time* was still thought of in terms of the analysis of equipment, and therefore functionally, came to be a privileged place of truth in 'The Origin of the Artwork'. At the same time, poetry, which was hardly mentioned in *Being and Time*, takes on a central role with his turn to Hölderlin in the mid-1930s. Heidegger's lectures on Hölderlin are discussed comprehensively in this chapter. It is argued that the central role of the work and of poiesis in Heidegger's writings in the 1930s can only be understood on the basis of his discussion of grounds. The question is, if philosophy cannot ground history – and the failure of *Being and Time* indicates for Heidegger that this is the case – then is there any ground at all? Heidegger argues that we experience grounds in the act of grounding, which he terms founding (*Stiftung*), but this means in poiesis. Through grounding, truth eventuates itself in the work. From these considerations, the questions arise as to whether kairos is to be understood poietically, and whether historicity can only be understood on the basis of poiesis.

The final chapter deals with the work in relation to politics. The question concerning Heidegger's political engagement in the years 1933–4 is discussed in the context of the relation of temporality and historicity. It is argued that although Heidegger's engagement with the Nazis went against the genuine tendency of his philosophical project, it is nonetheless motivated by an ambiguity in that project. In this context, an analysis of his Rectoral Address is attempted. The latter text is of philosophical interest because it shows Heidegger's understanding of the role of philosophy in a revolutionary political situation. It is shown that the philosophically significant claim of that speech is that thinking is historical, and that historical action is inevitably political. It is argued that Heidegger's failure lay in neglecting the intertwining of chronology with kairology – an intertwining which it has been shown can be worked out on the basis of his philosophical project – as it concerns political judgement and political responsibility. As such, the chapter ends with a section in the chronology of revolution.

1

Historicity and Temporality in *Being and Time*

I Kairos and chronos

I.1 History and time

> *'In what sense is time as passing definitive for history, when, on the other hand, we have said: history "emerges" [*'Inwiefern ist die Zeit als das Vergängliche bestimmend für die Geschichte, wo wir doch anderseits gesagt haben: Geschichte "entsteht"'*]?*
>
> (Heidegger, 1991, p. 62; cf. GA 38, p. 80; Heidegger, 2009, p. 100)[1]

Time for Heidegger lets things emerge as much as it lets them pass away. In terms of Heidegger's thought, the traditional account of time is one-sided in interpreting it as the happening of passing. Aristotle provides a classical example of this when he states: 'Time in and of itself is the origin of decay; time is the number of motion, changing movement however brings that which is towards its collapse.' (Aristotle, *Physics*, 221b1–2) Time has power over those who stand within it, and this power is traditionally understood as that of passing and decay.

But for something to pass, it must first emerge. Is this emergence a counter-power against time, or is there in time itself a power of emergence? This question needs to be posed in a more radical manner than the tradition from Aristotle to Hegel, which indentified time with passing (cf. Heidegger, 2010, p. 215; GA 21, p. 258) was able to do. Ought we inversely to identify time with emergence? To take this course, one must, along with Kant, understand the human not merely as an entity in time, but as herself temporal. This is so because what emerges does so through human agency; it does not belong to the cycles of nature but arises as something new. Such emergence, as the emerging of the new, occurs in the context of what is past. As such, passing is under the sway of time as emergence. This relation of emergence and passing defines historicity.[2]

We encounter historical time when time comes to appearance as that which is 'responsible' for emergence. We find this experience of time in human practice. In practice, something occurs at a particular time (which, as we shall see, is not to be understood as a now-point). Always at a certain time the coming towards us of the future (*Zu-kunft*, literally 'coming towards') occurs. In the arrival (*Ankunft*) of the future the 'new' comes upon us. In this sense, the future has power over us. Such an experience of time differs fundamentally from that which Aristotle assumes when he speaks of time as measure. We cannot reckon on this time; it is hardly possible to bring it into order through narration. Time appears as a power which we experience as at once 'chaotic' and ordering. I shall call the time experienced in this manner *kairological* time.

Heidegger himself used the term 'kairos' in the early 1920s, in particular in his interpretation of St. Paul.[3] For St. Paul, the time of the eschatological second coming of Christ cannot be calculated. One only knows that it will come suddenly. What is important is not to know the point in time at which the second coming will occur, but rather to be *prepared* for it. Such preparedness amounts to a heightened attentiveness. St. Paul commands his audience to be prepared as for a thief in the night, always alert, but without knowing when the time will come.[4] Time as kairos is the 'point in time' in which that which has no worldly correlation comes to appearance, but that time cannot be known either in advance nor – as it defeats all worldly knowledge – at the time of the coming itself. Only with the eyes of faith can it be encountered.

Kairos, though, is not 'contained' in the future; rather it is the moment (*Augenblick*) between past and future; it is the temporal dimension of decision. In this sense, for Paul, the decision of faith is already living in the kairos.[5]

While Paul gives kairos an apocalyptical sense, it is a term already to be found in Homer, for whom it means a part of the body, which is particularly vulnerable, due to a gap in the armour plating of a warrior. In battle, it is a particular success to hit that point with a spear; for the one wounded, however, it is a misfortune. (cf. Trédé, 1984, pp. xii–xiv; Homer, *Iliad*, IV, pp. v, 185) What we see here is an ambivalence in the kairos between success and misfortune, accentuated in Paul to apocalyptic proportions, as the second coming will bring judgement and hence again will be a moment of joy and of sorrow. When we generalize – as Heidegger does – the account of kairos to a kairology[6] of existence, and indeed of being, we find this fundamental ambivalence as marking time itself. Time as emergence and as passing away is no longer a mere backdrop of events, but a force or power both in which and through which being comes to appear for an entity which can experience it.[7] But, as with Paul, this experience is itself rooted in a decision, a decision which responds to the moment, and which acts in the moment in a manner appropriate to it. But if the moment cannot be reckoned with, its nature is essentially hidden also. The moment demands a decision which can bring about a new beginning, but can equally lead to catastrophe. This is the kairological situation of revolutionary action: to enter a new order, to take part in a revolution, can lead to happiness and peace, but also to disappointment and war. In the moment, mood swings between hope and despair, Angst and joy.[8] Kairos has these two aspects. It is the moment of decision, in which the standards of the past are no longer in force. The danger of catastrophe is correspondingly great. This danger arises from the discontinuity which characterizes the kairos. But out of such discontinuity history emerges; because, without such danger, the calling into question of the past would never arise and the future would never appear *as future*. Only when we see the past as contingent – as a past which had various possible futures and which could have been otherwise – is it possible to perceive historical change, that is, epochal movement. This possibility, however, assumes the experience of kairological time. In this sense, the kairos is not only historical; it forms the very possibility of history.

The term kairos is one which draws its meaning from both Greek and Christian sources.[9] In it, these two sources intersect in ways which are significant in the account which Heidegger gives. A parallel can be drawn between the straying from pre-Socratic thought in Platonism and that from early Christian faith experience in the Patristic period. Both of these 'forgettings' can be said to be characterized by the loss of the kairos – loss of a jarring discontinuity, in which the full questionableness of being comes to light. Furthermore, that which characterizes the Christian break is the introduction of a radical sense of discontinuity, a sense of an event through which an age emerges that does not simply repeat in a cyclical sense, as we find for example discussed at the beginning of the *Timaeus* (22b–26a), but rather marks the emergence of a new world.[10]

The term 'discontinuity' already suggests that time is for the most part not experienced kairologically, but rather as continuous. Time appears to us – and it is almost tautological to say so – as a chronological order. Nonetheless, in terms of its significance for chronology, continuity is a problem. Chronos tends towards dispersal which brings the being of time into question. 'Time' means the order of past, present and future. This order is, however, never complete, but must always be reproduced. In this respect, it is not accidental that 'chronicle' means the report concerning events during a particular temporal sequence.

Such a report is the subsequent placing in order of events. Chronology is the first step of narration. Further, chronology and causality are closely linked: every chronology assumes (in the widest sense) a causal connection between events. Conversely, it is not possible to speak of causal connections without reference to a certain chronology (at least in our normal ways of speaking). We will come back to this characteristic of chronology.

Chronology forms a continuity, a continuity which is suspended in the kairological moment. The relation between chronos and kairos is, nonetheless, problematic. While the occasion for the emergence of a kairological moment is to be found in the chronological – in time as continuity – the kairos also finds itself subsequently integrated into chronos. In kairos, there occurs a forgetting of chronos. Only the one who forgets chronos can act effectively in the kairos, because kairos suspends the authority of the existing normality. Nonetheless, it is prima facie to be expected that 'kairos' and 'chronos' in a certain sense must be the same, because both mean 'time'. This 'sameness' of chronos and kairos requires explanation. This question indicates the issue of an inner relation of ontology and ethics, which will need to be determined. This is so because – as will be shown, but which follows from what has been said already – kairos receives neither its explanation nor its justification from chronos. Must we conclude from this that action in the kairos is placed beyond ethics and political responsibility? Is an injustice done to the past – to our own past, the past of others, the past of our forbearers – in kairological action, insofar as forgetting is an injustice?[11]

To this point, kairos has been spoken of exclusively as a phenomenon of historicity. It is not, however, the case that only historical events can be characterized in terms of the kairos. Certainly the kairos is both an intensification of the everyday and a stepping outside of the everyday. Such an event can, nevertheless, occur within an already established order while that order is preserved, and is so in part thanks to that event. An example of this is religious initiation. In this case, a person is elevated in a moment – a moment which can last minutes or months – from the stage of childhood to that of adulthood. Externally nothing need change in the appearance of that person, but he is nonetheless recognized otherwise. In the context of religious life this change can only occur through a discontinuity.[12] The sacraments of the Christian churches have kairological characteristics, insofar as in them a transformation takes place and the world of the believers becomes other. What this means is that the kairos is not to be confined only to world-changing events, but rather is constitutive of belonging to a community, indeed to a historical people.

If, however, we understand historicity in such a wide way, do we not find in the end that the very distinction between temporality and historicity is an abstract one? We can pose the question this way: In every predication of time there is an objective as well as a subjective genitive. Psychological time, for example, means both the time which the psyche experiences (objective genitive) and the structure of this experience (subjective genitive). Similarly, cosmological time means the time as the measure of the movement of the cosmos (objective genitive), but also this movement as temporally ordered (subjective genitive).[13] 'Historical time' is both the time of history and also time as a historical time. On the one hand, the historical is temporal. If there was no time, there would be no change, and without change there would be no history. On the other hand, history is a coursing of time; it is not only the sequence of events in time, but is also the coursing of time itself understood as a principle of human destiny. Time is, hence, also essentially historical.

There are two ways of avoiding this consequence, both of which, however, Heidegger rejects in *Being and Time*. Firstly, one could deny the universality of the predicate 'time'. One could thus say that time is historical only for certain entities, but that time as such is not historical as there are entities which are not historical. This option Heidegger would deny, because it would amount to an understanding of the ontological account of time as an abstract generalization from the ontic constitution of certain entities. Secondly, one could give a genealogical account of historicity. In such an account, one could claim that time is historical only if the human being is historical, but that the human being was not always historical.[14] Heidegger claims against this that historicity is derivative from temporality of which it is a concrete development. This points to an ambiguity at the heart of *Being and Time*.[15]

In *Being and Time*, Heidegger attempts, on the one hand, to ground history; on the other hand, his whole project is one of repeating history. The work is structured through such a repetition or retrieval.[16] The second part is explicitly a repetition of the first, and the first part is a repetition of the history of ontology.[17] Heidegger states: 'We call authentic having-been "*repetition*" (*Wiederholung*)'. (BT, p. 311; H, p. 339) Authentic Dasein repeats past possibilities. If *Being and Time* is a repetition, we must ask, what is repeated? *Being and Time* repeats certain specific possibilities – namely, the having-been of the history of ontology. For this reason, the structure of Dasein, which is revealed in this repetition, has a historical character.

But the question concerning the relationship of temporality and historicity in *Being and Time* refers to a deeper basis in the systematics of that work. A hint of this can be found in a short remark Heidegger makes concerning the ordering principle of that work. In the context of a critique of ethnology, he objects to the 'syncretistic comparison and classification of everything', and then goes on:

> The genuine principle of order has its own content which is never found by ordering, but is rather already presupposed in ordering. Thus the explicit idea of world as such is a prerequisite for the order of world images. And if 'world' itself is constitutive of Dasein, the conceptual development of the phenomenon of world requires an insight into the fundamental structures of Dasein. (BT, p. 48; H, p. 52).

Here, we can see that the whole structure of *Being and Time* assumes an ordering principle. The method of *Being and Time*, always beginning with the immediate, and from there to strive for the origin is a way to the ordering principle. When it is the aim of *Being and Time* to pose the question concerning being, its inner movement is towards that order of being which is already presupposed in those entities which it is investigating. There is an inner circularity here – which Heidegger makes explicit in discussing the hermeneutical circle in § 32 of *Being and Time* – which guides the investigation from the beginning.[18] This inner circularity is confirmed in the movement of repetition or retrieval: in repetition, the ordering principle appears because the repeated is allowed to appear in the possibility of its own appearance. The task of philosophical thought is not to construct or even reconstruct that order but to think the apparent in the event of its own appearance. It is that event of appearance – the world – that is presupposed in the ordering of entities.[19]

I.2 The moment of vision (*Augenblick*)[20]

Action in the moment of vision (*Augenblick*) is a repetition because the moment is unique and cannot be generalized. Every attempt to act in the moment through the application of a rule will fail because the singularity of the moment suspends time understood as a quantitative and indifferent now-sequence. This does not mean that the moment is without all relation. When Kierkegaard, the thinker of the moment, speaks of the 'contemporary disciples' and the 'disciples at second hand' (cf. Kierkegaard, 1962, pp. 68–137) or of Adam and other humanity (Kierkegaard, 1980, p. 81), he displays an acute sensitivity for historical difference. There is an ambiguity in this difference. On the one hand, every moment appears as the first: it is a qualitative leap as if without precedent. On the other hand, between every such leap there is a quantitatively measurable distance, the relation over such distance is constitutive of history. (cf. Kierkegaard, 1980, pp. 30–1) The underlying issue here is how to understand the relation of continuity and discontinuity as constitutive of movement. If in movement – understood in the wider sense of '*kinesis*' – change occurs, then there must be discontinuity, but there is movement only if continuity between present and past, and present and future is possible. (cf. see Gadamer, 1993; 1999) Kierkegaard wrote in his 'Notebooks' that the most important problem was that of movement. (cf. Kierkegaard, 1980, p. 242) The moment can only be of significance if movement and change are understood non-metaphysically. Metaphysically understood change and movement assume the constancy of that which changes – the constancy of an underlying substance.[21] This assumption is also at the basis of that which Kierkegaard terms the 'Socratic situation' of knowing already that which one strives to know. (cf. Kierkegaard, 1962, pp. 11–12) In the latter situation, the moment has no significance. Rather, there is there a speculative structure, which is directed at the past. The teacher reminds the students of the eternal ideas. The 'moment' of recall supplies only the occasion for the confirmation of the pre-existing truth; it has nothing to do with the truth as such. That which remains always constant does not change. In the Socratic moment, history is separated from the eternal. The historical place of Socrates and his students has no relevance to the truth. Truth lies in the ideas which are eternal and earlier than any temporally early point. For Kierkegaard, however, Christianity breaks with the Socratic situation through its understanding of the moment as the intersection of the eternal and time. (Kierkegaard, 1962, p. 79) The eternal does not lie behind the Christian, accessible through recollection, but rather lies ahead of her, accomplished through repetition.

It is clear that these reflections of Kierkegaard were influential on Heidegger.[22] The latter's concept of the moment can be understood as an attempt to think Kierkegaard's critique of the Socratic moment through to its conclusion. As with Kierkegaard, Heidegger is attempting to understand a way of knowing which is not directed towards an already known, but rather forwards towards that which can emerges anew. He refuses, however, Kierkegaard's account of such knowledge as faith. Rather, he is seeking a mode of thinking which is neutral with respect to Christianity. Furthermore, he seeks to break with Kierkegaard's attempt to understand the moment in terms of an intersection of eternity and time. Eternity for Heidegger is rooted in an interpretation of being as presence, according to which genuine being is that which is never not there, what is always there: *ἀεί ὄν*, always being, the putative root of the Greek word for eternity, *αἰών*. Such an understanding of being makes it impossible to think of human beings as historical. (cf. Heidegger, 2007b, p. 273; KV, p. 176) In the Platonic tradition, the eternal is understood as the perfect cyclical movement of the heavenly spheres. (cf. Plato, *Timaeus* 33b4–9 and 37d3–e2) Finite temporality cannot be thought on that basis because the cycle is infinite. The point here is not simply one of inversion, of claiming a linearity of time in contrast to the cyclicality of the eternal. It is rather the case that Heidegger seeks to avoid any understanding of the moment, and hence of kairological temporality, as that of a self-circling recurrence.

Heidegger's own path to the moment is neither theological nor philosophical, but rather pre-scientific, rooted in the everyday. (Heidegger, 2007, p. 201; GA 64, p. 108) This does not mean taking over the conceptuality of the everyday. The everyday understanding, according to Heidegger, forgets the difference between Dasein and non-Dasein-like entities. Dasein is that entity which is concerned with its own being. Its form of being is one of existence, not of the objectively present (*vorhandene*). Existence has the structure of care (*Sorge*). The structure of care is temporal and this leads Heidegger to a consideration of the moment. The structure of care he defines as 'Ahead-of-itself-already-being-in (a world) as being-together-with (entities encountered within the world)'. (BT, p. 300; H, p. 327 [translation modified]) The temporal meaning of this structure is already formally indicated by the terms 'ahead' and 'already'. Such an indication is, Heidegger tells us, here lacking for the third element of care in the above definition, namely that of the fallenness of the being-together-with, that entanglement in things with which Dasein is concerned. What is hinted at here is the '*making present*, as the *primary* basis for the *falling into entanglement* with the things at hand and the objectively present of our concerns, remains *included* in the future and in having-been (*Gewesenheit*) in the mode of primordial temporality.' (BT, p. 301; H, p. 328 [emphasis in original; translation modified]).

The question which arises here is whether, given this account of care, the present can be a distinct mode of temporality. Is the present not simply absorbed into the having-been and the future? The importance of this question for the present work is that Dasein can only act in an authentic manner through a fetching itself out of fallenness in the moment. The question can be formulated in this way: how can Dasein break out of the chronologically-constituted present to act in the moment?[23]

The moment is a breaking out of fallenness. In temporal terms, we could understand this as a 'dechronologization' of the present. This is the authentic present. (cf. BT, p. 311; H, p. 338) For the most part, Dasein is inauthentic. In terms of the present, this means that Dasein loses itself in its most immediate concerns. To emerge out of this is, for Dasein, to bring itself back to its authentic present. The present from which Dasein brings itself back remains, however, indistinct. It is contained in the future and the having-been. Although the ontological basis of fallenness lies, for Heidegger, in the present, fallen Dasein does not know the present, because it is dispersed in the having-been and the future. On the other hand, Dasein uses such temporal terms as 'now', 'on that former occasion (*damals*)' and 'then', and as such implicitly knows the present as the reference point for setting dates and times. (cf. BT, pp. 373–4; H, pp. 406–7). If Dasein had no awareness of such a present, it would not have any knowledge of time, as only an entity which has such awareness of the present can know the difference between having-been and future. In this sense, though, the present remains simply the place of the difference between having-been and future. As such, it cannot form the basis of concern for the objectively present and the handy – things as equipment or as objects of perception and knowledge. For instance, it is only in the present that a piece of equipment can be 'to hand'. The present of this handiness is the present of action.[24] Its presence makes this action possible. As Heidegger states: 'letting *what presences* in the surrounding world be encountered in action, is possible only in making that being *present*.' (BT, p. 300; H, p. 326 [emphasis in original])

The present has, in any case, the character of transition. It disappears into the having-been. In this disappearance, it remains under the power of the having-been. At the same time the new, the novel, emerges in the present. Such emergence is a coming to be and as such remains under the power of the future – the 'to come'. The present of action is the tension between these two forces. Its own power can only consist in its resistance to the future and the having-been.[25] Furthermore, only in this resistance can there be a destructuring (*Destruktion*) of the history of ontology, which is the task of *Being and Time*, because destructuring concerns the today (BT, p. 20, H, pp. 22–3) Only through the force of this resistance is it possible for chronological time to be suspended and for the moment of vision to occur.

The moment of vision has two meanings in *Being and Time*: the authentic present (cf. BT, p. 311; H, p. 338) and the unity of authentic time generally (BT, pp. 301–2; H, p. 328). Similarly, the kairos is not simply the authentic present, but authentic temporality as such. This ambiguity of the present is due to its peculiar place as what appears the most real in time and which slips away between past and future. In this way, the moment reflects the ambiguous place of the inauthentic present, as, on the one hand, co-original with the having-been and the future and, on the other, disappearing in those two temporal modi. A passage from his 'Kassel Lectures' gives some insight into the actual basis of this ambiguity. There, he states that the present lies in action. What is remarkable here, according to Heidegger, is that

> the past becomes alive and the present disappears whenever we act in the direction of the future. Those who act authentically, live out of the future (*Eigentlich handeln die, die aus der Zukunft leben*), and can also live out of the past, while the present takes care of itself (Heidegger, 2007b, p. 266; KV, p. 169)

Dasein acts authentically in the present, but precisely in action the present remains hidden. There is in the present no reflective consciousness of the present; the agent cannot observe his present self, because action demands a losing of the self in giving oneself over to the future and the having-been. This should not, however, disguise the fact that action always happens in the present and that it carries within itself the possibility of resistance against the having-been and the future.

The moment can only emerge out of this resistance, and as such out of a unity in conflict (*polemos*). Such unity has the character of movement. According to Heidegger, the movement of existence is different from the motion of the objectively present. (BT, p. 344; H, p. 374–5) As such, Dasein has a particular movement, which Heidegger terms the 'stretched along and stretching itself along', the 'occurrence (*Geschehen*)' of Dasein.[26] This stretching is between birth and death, Heidegger says, following Dilthey. It is important to recognize the temporal presupposition of this characterization. In his discussion of Aristotle, Heidegger stresses the stretching of the now. The now has the character of transition (*Übergang*). (cf. Heidegger, 1988, p. 248; GA 24, p. 352). Only when we see this, according to Heidegger, can we understand Aristotle's account of movement (*kinesis*). Heidegger's whole analysis of temporality in *Being and Time* centres around the attempt to show that this Aristotelian now (*nun*) is derivative. The stretching of the now gives us a sense of the character of movement, but only in a derivative way. That which characterizes the movement of Dasein remains hidden because it is not based on the stretching of the now, but rather of the moment.

Heidegger's aim in the exposition of historicity is 'to lead us to the ontological enigma of the movement of the occurrence in general.' (BT, p. 355; H, p. 389) Movement here is precisely the movement of the moment, which Kierkegaard understood as the movement of becoming. (Kierkegaard, 1962, pp. 91–3) As Kierkegaard points out, Plato was already aware of the mystery of such movement. That which Kierkegaard called the 'moment', is in Plato the *ἐξαίφνης*, the 'sudden': that which is not contained in the normal sequence of events. It is neither moved nor still, neither being nor non-being. It is a transition and as such something *out of* the ordinary.[27]

But if the moment is to be understood in this way, how can it also be conceived as forming the unity of time? In order to make the issue clear, but also to find a way of resolving it, we can take a hint from Georg Simmel's essay 'The Adventure' ('*Das Abenteuer*').[28] There is in adventure, as described by Simmel, an analogy to the moment of vision. An adventure falls out of the context of life. (Simmel, 1997, p. 222) The adventure by its nature does not belong in the continuity of the rest of the person's life. It does, nonetheless, return to the movement of the life, but as a foreign body – like a dream. Indeed, for Simmel, the adventure takes on the hue of a dream. (Simmel, 1997, p. 222) In adventure' there is from the beginning a strangeness, which has the dream-like quality of being out of the normal course of things. (cf. Simmel, 1997, pp. 223, 229)

In a dream, the world is otherwise than it normally is. The normal conditions of the world, the boundaries of normal human possibilities – boundaries which protect us from our deepest fears – are no longer present. We find ourselves in a different world. And yet, while the world of the dream-state is not real, it nonetheless has relations with the world of our waking state. This relation is, however, not a direct one. The dream-world is cut off from the waking world, but the dream can be remembered, can be narrated, and can as such be brought into the inter-subjective space of the waking world. Furthermore, the dream can give us insights concerning our everyday lives. A rationalistic tendency both in philosophy and in everyday life seeks to dismiss the dream as much as the adventure as insignificant for life. As Heidegger shows, we judge the unusual and the extraordinary by the measure of the ordinary and condemn or disregard it accordingly. The dream and the adventure are judged according to the standard of the waking and normal world.[29] According to such a view, neither the dream nor the adventure can have any real significance. That which does not correspond to the normal expectations, experiences and perceptions of the everyday, is for the most part suppressed, that is, pushed out of the sphere of significance and almost totally out of all recognition. But this is never totally the case, because the foreignness of the dream and of the adventure lies in the manner in which it accentuates that which, in the normal course of life, lies beyond all meaning and significance. Simmel speaks in this context of a principle of accentuation (*Prinzip der Akzentuierung*), which accentuates the hue, temperature and rhythm of the process of life at the expense of the content of the life lived, i.e. of that which can be propositionally stated. (Simmel, 1997, pp. 229–31) This means that the strangeness of the adventure and the dream does not indicate their irrelevance for life in its connectedness (*Zusammenhang des Lebens*), but rather that through their discontinuity with that life they allow the connectedness of life to appear as such – not as content but as movement.

Discontinuity so understood constitutes history. The being of historicity is constituted by the occurrence, at the basis of which lies such discontinuity. (BT, p. 17; H, pp. 19–20). Historicity is at the same time the basis of any possible historical understanding, and indeed of history as an academic discipline (*Wissenschaft*). (cf. BT, p. 305; H, p. 332.) As such discontinuity, appears to be the condition of possibility of continuity. Otherwise stated, chronos presupposes kairos.

In order to understand how kairos might be the condition of possibility of chronos, we need to investigate what is meant by 'possibility'. It is crucial in understanding Heidegger generally, and the themes of this book in particular, to follow the shift in meaning of possibility which Heidegger effects. Kierkegaard and Bergson are precursors of Heidegger in this respect. Each of these three philosophers share a basic, fundamental thought in common: if we are to understand possibility on the basis of freedom, then it can no longer be thought of as a realm of present options which can be chosen, but rather must be thought of in terms of a future which is neither present nor pre-formed in the present. This shift in the meaning of possibility arises out of a critique of Aristotle. For the latter, actuality has priority over possibility. (cf. Aristotle, *Metaphysics*, 1049b5) Accordingly, possibility is already contained in an entity, is the possibility of that entity, not yet fulfilled. This 'not yet' indicates a lack – the lack of actuality – in possibility. Heidegger, when he states that possibility is higher than actuality (BT 34; H, 38), is deliberately rejecting this thesis. Heidegger is not simply inverting Aristotle here, but is calling Aristotle's concept of possibility into question. (cf. Figal, 1988, p. 91) Strictly speaking, Heidegger is no longer understanding possibility as a modal category. A modal category is a category of something which is in the state of being expressed by that modal category – possible, actual or necessary. The entity is determined in terms of these categories; without them, its being remains undetermined. Possibility in such a view is 'not yet'. There is the possibility that the puppy I saw yesterday on my walk will one day become an old dog. This possibility is objectively present (*vorhanden*), but is not yet actual, nor is it necessary. However, 'Dasein is not something objectively present which then has in addition the ability to do something, but is rather primarily being-possible. Dasein is always what it can be and how it is its possibility.' (BT, p. 134; H, p. 143)[30] Furthermore, Heidegger goes on to say, 'Dasein is the possibility of being free *for* its ownmost potentiality-of-being.' (BT, p. 135; H, p. 144) Understanding (*Verstehen*) is the being of such potentiality-of-being.

We can only sensibly speak of understanding if we remain within the realm of possibility. Understanding presupposes not-knowing. Understanding as such has the character of projection (*Entwurf*). Richard Kearney expressed the difference between possibility and potentiality-to-be in this way: 'while the "possibility (*Möglichkeit*)" is the projection of Dasein, the potentiality-to-be (*Seinskönnen*) is Dasein itself which projects. The "potentiality to be" is that which makes possible all projection towards a possibility'. (Kearney, 1984, p. 122) In other words, possibility is not a modality of Dasein because Dasein has the mode of being of possibility in the sense of making possible. The making possible is the being of Dasein as an entity which 'understands (*Verstehen*)'. Understanding is primarily directed towards the future. (BT, p. 310; H, p. 337) As such Dasein cannot reflect upon itself as it would on an objectively present thing. Its 'knowledge' of its own potentiality-of-being arises only through the possibilities, which are projected on the basis of its thrownness. If that were not the case, Dasein would understand itself only in terms of already present possibilities. But if, as Heidegger states, Dasein is delivered over to the possibility 'of first finding itself again in its possibilities' (BT, p. 135; H, p. 144), then Dasein misunderstands itself to the extent to which it takes itself to be objectively present.

'As projecting, understanding is the mode of being of Dasein in which it *is* its possibilities as possibilities.' (BT, p. 136; H, p. 145) Projection is a movement of Dasein, which is irreducible to the movement of thrownness. In other words, the movement of projection brings possibilities into play which are not simply already there, but which Dasein encounters in the moment. (cf. BT, pp. 311–12; H, pp. 338–9) As with Bergson, what is at issue here is an ambiguity regarding 'possibility'. We can say with Aristotle that nothing happens which was not previously possible. But Bergson in response gives a counter-example (cf. Bergson, 1968, p. 22): before a musician composes a symphony, that piece of music is already possible only in the negative sense, that nothing hinders its coming to be. If we were to speak in a positive sense, we would have to say that, prior to its composition, the piece of music was already contained in the composer's apperception. But, for Bergson, it is absurd to speak of an artwork in this way. If we do not see this ambiguity in possibility, we are unable to understand freedom or novelty, according to Bergson. If, following Aristotle, we conceive possibility only as real (non-actualized) possibility, then in the case of Bergson's example, we can no longer speak of possibility. Heidegger, however, speaks of possibility as making possible (*Ermöglichung*). The symphony is not an already present possibility, which can at some time be actualized; rather, in the process of composing, the making possible of the piece occurs, an opening up of a structure, in which a new future is initiated. The novelty of this new future cannot be found in the realm of the actual.[31]

Possibility as possibility cannot be made thematic for Heidegger. Dasein, which projects itself according to its possibilities, does not have a theoretical relationship to them. Rather, the possibilities of Dasein are what it itself is. Heidegger discusses this with respect to the most radical possibility of Dasein, the possibility of impossibility, death. 'Terminologically, we shall formulate this being toward possibility *"anticipation" of this possibility*.' (BT, p. 242; H, p. 262) Such anticipation is more literally a 'going-out-ahead-of-oneself (*vor-laufen*)', and in respect to this ultimate possibility is a movement of the authentic being towards death. Death as a possibility makes things strange. It cannot be understood as the possibility of an actuality, as it is precisely that which can never be actual. It is not an event in the world; it is rather Dasein itself insofar as it is finite.

The analysis of possibility goes hand in hand with the analysis of death. Dasein relates to itself as possibility because it is temporal. Objective presence (*Vorhandenheit*) is an abstraction of time, which when applied to Dasein disguises its temporality. The temporality of Dasein can only be understood as possibility, because the moment is futural, and as such a break with the past and requiring a leap. The temporality of Dasein is, therefore, that of change and, finally, of rupture. Such discontinuity is kairological and radically historical. History, however, is the forming of continuity. It is the forming of binding connections with the past. History requires narration; it must have a unity; it may even have a teleology, if not an eschatology.

I.3 History and historiography

My interest here is not so much in the question concerning the relation between science (of history) and ontology (on the latter topic cf. Hoy, 1978), as it is in the underlying difference in time. The principle of order mentioned previously is evident again here. Historiography (*Historie*), that is the science of history, history as an academic discipline, has history (*Geschichte*) as its object, the meaning of which historiography already assumes.

We encounter history in the houses in which we live; on the streets, where we step in the footsteps of earlier generations; in the ruins, where we can almost hear the echoes of lost worlds; through stories, at monuments, on the hills and in the valleys marked by the events and people of earlier times. History as an academic discipline plays in each of these cases for us a crucial role, as it sets the boundaries of what is to be considered 'historical' or of 'historical interest'. In historiography, history is treated as the content of a form of knowledge. This knowledge is theoretical. Its goal is to set out the truth of history. But what kind of truth is that? It is in any case a mediated truth: the historian has no direct access to his object, the past. Every access he has is mediated through documents, architecture, witnesses of different kinds.

With respect to historiography, Heidegger says: 'The idea of historiography as a science implies that it has grasped the *disclosure* of historical entities as its own task.' (BT, p. 359; H, p. 393 [translation modified]) The disclosure in question here is of having-been Dasein in its having-been possibilities. Historiography concerns facts, but these facts are constituted by Dasein's 'resolute self-projection upon a chosen potentiality-of-being'. (BT, p. 360; H, p. 394) The converse side of this is that, while historiography indeed discloses the 'still force of the possible', it does so precisely not as possibility, but as fact. This means that historiography remains on the level of chronology, that of the temporal sequence of 'facts', which arise out of the force of the possible through repetition, i.e. historicity. David Hoy failed to see this, and hence is led to conclude that, for Heidegger, the historian must be concerned not with facts, but with possibilities. (cf. Hoy, 1978, pp. 349–50) It is indeed the case that Heidegger states the *theme* of historiography to be the possibility of having-been existence. The theme is, however, the 'horizon' of a projection, which holds a particular region of entities. (cf. BT, p. 359; H, p. 393) Within this horizon are the objects of the specific sciences, the entities as present-at-hand. Historiography does not disclose its theme in its truth. It remains tied to its objects. Without understanding this difference between theme and object, Heidegger's attempt to transcend historiography and chronology must remain obscure.

Facts are clearly the objects of historiography. But the thinker in the background of these analyses is Nietzsche. (cf. Taminiaux, 1991, pp. 175–90) It is not by accident that Heidegger ends the section from *Being and Time* to which I have been referring (§ 76) with a discussion of Nietzsche's threefold differentiation of history (*Historie*). For Nietzsche, history as an academic discipline cannot be judged simply by academic, scientific standards. History as an academic discipline endangers us because it makes us blind for the kairological possibilities of the moment. In each of the three forms of history which Nietzsche discusses – monumental, antiquarian and critical – there is a tendency to level off history to a mere chronological continuity. The danger here is that of neglecting the possibility of a truly lively existence. In order to avoid this danger we must possess the strength to forget the past, that is to live in equal measure historically and unhistorically. (cf. Nietzsche, 1997, p. 63) If the human being acts in that way, she forgets chronology and acts in the uniqueness of the moment.[32]

For Heidegger, the chronological (historiography) is founded on the disclosure of historical being. This raises a number of questions, in particular: how does this disclosure of historical possibility differ from fundamental ontology? How, in other words, do we achieve access to having-been Dasein in the first place?

II Historicity and kairological temporality

Kairos is the time of transition. Transition is a movement, one which brings change, even radical change. In the kairos, what is of concern is not the mere observation of phenomena, but rather self-movement in a world of change. As we know from Aristotle, the world of change concerns practice, and specifically knowledge concerning practice.[33] For his reason, we must investigate kairos within practice. We have, however, already said that the kairos is historical. In Aristotle, we find no discussion of the historicity of human Dasein. His account of the kairos is one of a rhetorical-political sphere.[34] The fundamental ontological implications of this account – and of his account of practical truth generally – are not pursued by Aristotle. It is for this reason that Heidegger must go beyond Aristotle in developing a concept of practice. He does not do this explicitly. We need to interrogate his texts indirectly, in order to recognize this shift in the concept of practice. Beginning with Heidegger's account of historicity, I will try to clarify the concept of practice underlying it.

II.1 Historicity, practice, birth

In section 74 of *Being and Time,* Heidegger makes clear how we should understand historicity in relation to temporality: '[T]he interpretation of the historicity of Dasein turns out to be basically just a more concrete working out of temporality.' (BT, p. 350; H, p. 382). Temporality is more original than historicity. If we understand Dasein's temporality in terms of the structure of care, as Heidegger does, think the latter as being towards death, and understand death as that to which Dasein stands only as a singular being, then we must ask how historicity can be understood on this basis. By reading *Being and Time* in reverse – that is, from the sections on historicity backwards – I will attempt to answer this question.

'The thesis of the historicity of Dasein does not say that the worldless subject is historical, but that what is historical is the entity that exists as being-in-the-world.' (BT, p. 355; H, p. 388) Furthermore, he states that the historicity of Dasein is fundamentally that of the world. Only because Dasein has a world are there historical things. The 'antique', which is not yet past, is historical because the world, in which it belonged to a connectedness of equipment, is no longer. (BT, p. 348; H, p. 380.)

Although world is a central concept for Heidegger, my specific concern here is with the *historical* world of Dasein. It is the world of occurrences (*Geschehnisse*), the 'specific movement of stretched along and stretching itself along' (BT, p. 344; H, p. 375), i.e. the movement within which there occurs a connectedness of the disclosing stretching of Dasein between birth and death. Heidegger admits at this point that 'the "connectedness (*Zusammenhang*) of life" in which, after all, Dasein constantly somehow holds itself was overlooked in our analysis of being-a-whole.' (BT, p. 342; H, p. 373) Although Heidegger had discussed at length the one end of Dasein – death – he had neglected the other – birth. Hence, Heidegger states, his analysis to this point has been one-sided. This admission is difficult to reconcile with the claim, already noted and often repeated in the historicity sections of *Being and Time*, that historicity is only a working-out of temporality. There must be a reason why Heidegger only discussed birth in the wake of his analysis of temporality and in the context of his historicity discussions: if temporality is Heidegger's answer to the question of the totality of Dasein, why did he not analyse birth within the discussion of temporality?

I will leave this question hanging for the moment, and instead begin with the question of the meaning of birth for Heidegger and its relation to historicity. Clearly birth plays a different role from death. For example, Heidegger states that '"birth" is *taken up into its existence* in coming back from the possibility of death, the possibility not to be by-passed'. (BT, p. 357; H, p. 391 [emphasis in original]) Death cannot be by-passed (un*über*holbar), birth is taken up into (*ein*geholt); the other ending – birth – can be brought into existence as a having-been possibility. Birth is, in this sense, heritage [*Erbe*]. Death, on the other hand, is annihilation. It transcends all inheritance. Birth involves essentially a taking up of community; in death, the community withdraws from Dasein. (cf. O'Byrne, 2010, pp. 29–35)

Despite these differences, death and birth have in common that they are ends, and Dasein, because it lies between these two ends (and indeed *is* this between), is finite. Dasein's fate arises from this finitude. Along with 'birth', 'occurrence', and 'destiny' (*Geschick*), Heidegger employs the term 'fate (*Schicksal*)' here systematically for the first time. He introduces the term in relation to death, but the decisive context is finitude. As Heidegger states:

> The finitude of existence when seized upon tears one back out of the endless multiplicity of possibilities offering themselves nearest by – those of comfort, shirking and taking things easy – and brings Dasein into the simplicity of its fate (*Schicksal*). This is how we designate primordial occurrence (*Geschehen*) of Dasein, which lies in authentic resoluteness in which it hands itself down to itself, free for death, in a possibility that it has inherited and yet has chosen. (BT, p. 351; H, p. 384)

Although Heidegger does not explicitly mention birth here, it is nonetheless taken into existence through its handing down of inheritance. Only through birth does Dasein have inherited possibility.

'The fateful destiny of Dasein in and with its "generation" goes to make up the full authentic occurrence of Dasein.' (BT, p. 352; H, pp. 384–5) To the extent to which Heidegger introduces the concept of 'generation', he alludes to Dilthey.[35] For Dilthey, 'generation' is a phenomenon of contemporeneity. To live with others in a generation means to be contemporaneous with them. Although Dilthey does speak in this context of the mean (*Durchschnittlichkeit*), 'generation' is not a concept of measure. Rather, for Dilthey, there is a narrower and wider concept of generation. The wider concept is that of neutral contemporaneity, which amounts to about 30 years. The narrower concept refers to a circle of individuals, who are bound together by their common experiences of the same events and changes. The First World War is a good example here: those who were 16 years old in 1914 did not belong to the generation of those who were 18 in that year, because they were not of age to enter the army. Heidegger seems to have this narrower concept in mind, because the occurrence of Dasein is determined by the fateful destiny of Dasein in and with its generation. (cf. BT, p. 351f; H, p. 384) As such, 'generation' has a kairological meaning. The latter, however, cannot be separated from its chronological meaning because each generation relates to another through a chronological sequence. Birth conditions this relation of the chronological and the kairological.

'Generation' has no meaning in abstraction from birth. Generation forms the connection to the historical process. This means that Dasein as member of a community of, as Heidegger also says, a people ('*Volk*'), is already connected with a past and implicitly with a future. To the extent to which Dasein experiences other generations, past and future are experienced as they relate to one another. Generation is not simply a biological phenomenon: it is the handing over (tradition, *Überlieferung*) of past to future, which makes a community possible. (cf. Held, 1996)

Heidegger employs the concept of generation only twice in *Being and Time*. The first time is in section 6, where he presents the programme of a destructuring (*Destruktion*) of the history of ontology. In that context, Heidegger remarks that *Dasein's* own past is at the same time the past of its whole generation. That means that historicity has reality only for a community, or rather, for Dasein, as a member of a community. Later, as mentioned, in the historicity sections, he returns to this concept in relation to the destiny of a community or people. Hence, both at the beginning and end of his magnum opus, Heidegger emphasizes the historical, in the sense of generational, being of Dasein and does so in a manner which ties this historical being to the project of destructuring. Dasein can approach the history of ontology only from the place of its own generation. This does not mean that it becomes simply a mouthpiece of its generation; on the contrary, as a member of a generation, it is called upon to destructure its own entanglement in the ontological understanding of its tradition. (cf. BT, pp. 18–19; H, p. 21)

There is no sentence in *Being and Time* in which Heidegger speaks about the connection between birth, generation and destiny. Nonetheless, these concepts in their connectedness form the basis of an understanding of historicity which Heidegger had not yet fully seen in that work. Rather, because he one-sidedly considers the finitude of Dasein as being towards death, he tends to identify finitude with totality. But with respect to birth as another form of end, finitude must be thought of differently. We can state the thesis in this way: the finitude of Dasein understood on the basis of birth is not the finitude of totality, but of loss and novelty; this is so because with birth Dasein begins – with its generation – and in every beginning, but especially that of birth, there stands the loss of parting – taking leave of the generation(s) from which it comes – and by the same token the novelty of initiation.

Heidegger understands natal Dasein as itself being towards death. As born, Dasein is already destined to die. And through birth, Dasein is thrown into a particular people. This people has a history, which is carried on through each generation. This is what we call 'heritage'. Through such heritage, there is destiny as the 'occurrence of the community of a people'. A destiny is a being with one another. The word '*Geschick*' in German – which we translate as 'destiny' – comes, like '*Schicksal* (fate)', from the verb '*schicken*'. Today, the latter verb is synonymous in ordinary German with '*senden*' (to send). Etymologically, however, *schicken* means 'to order, to prepare, to put in work'. (cf. *Grimmisches Wörterbuch*, vol. 14, columns 2644–57) It means to make something ready to reach a goal, or to make something appropriate for some end. It is to make it fit for something. Understood in these terms, we can say that the destiny of a people is its ability or preparedness to play an appropriate role in a particular historical situation. Such destiny comes from the past and reaches into the future.

It is similar with birth. Heidegger says in this context: 'understood existentially, birth is never something past in the sense of what is no longer objectively present, and death is just as far from having the kind of being of something outstanding that is not yet objectively present but will come.' (BT, p. 343; H, p. 374) Birth is not past, but rather 'having-been'. To be 'having-been' means to be historical. (BT, p. 359; H, p. 393) The fact that Dasein is stretched does not mean that its past is no more; at most, it is hidden. As such, the past is there as possibility, which can be repeated. As Dastur puts it, authentic history 'is the return of a possible having-been'. (Dastur, 1998, p. 46) Birth is repeated as possibility. The possibility of birth is that of entry into a certain history of a certain community and of a certain people, which is the possibility to be historical.

Birth, then, is to be thought of as a transition; it is a movement between two orders. Birth is an overturning, which can be understood as at once a leap and a development. (cf. Saner, 1976) As such, birth is an exemplary coincidence of chronos and kairos. The turning over of birth is qualitative as much as quantitative. It is a leap from the foetal environment to a cultural world. This leap occurs in the moment of birth, which is irreducible to quantitative determination. As a moment of transition, it is a threshold (*Schwelle*), to use a term from Walter Benjamin. (Waldenfels, 1987, pp. 28–9) Decisive in this leap is openness. The newborn has, in contrast to its foetal existence, the possibility to project itself and to have a world. With birth, a world is opened up, an order of relations, which lies before the newborn as pure possibility. The leap has something abysmal about it. Being born occurs neither in the foetal environment, nor in the world of Dasein. The violence of birth is the violence of transition, which because it has no ground (*Ab-grund*), because there is no synthesis between the foetal environment and the world of Dasein, demands a leap without guarantee. (Waldenfels, 1987, pp. 22–4)

But seen from another perspective, birth is not a leap, but only a step in a developmental process. The newborn child has the same identity as the foetus.[36] There is evidently a continuity – not only biological, but also in terms of identity – which encompasses birth. Furthermore, the newborn is in many ways closer to the foetus than the more developed person; like the foetus, she lacks many of those characteristics which we associate with human beings: linguistic ability, capacity for independent living, creative ability. Certainly birth brings changes, but these are simply parts of the normal development of a human being. There is here a clear chronology.

But such a chronological understanding misses a crucial point. Understood in chronological terms, birth appears as one moment – albeit with special significance[37] – amongst others in a sequence of events and situations. Externally understood, this is the case. But the movement which occurs in birth is irreducible to external observation. The transition which occurs there is, strictly speaking, beyond all experience. Nobody experiences birth, because birth makes experience possible. The experience of birth could only be described with the categories of the worldly conditions of experience. But experience of the world is precisely what birth makes possible. It is for this reason that Hans Saner speaks of the leap of birth as having a transcendental significance. (Saner, 1976, p. 150) There is in birth a moment of transition, which can neither be explained or justified from what went before or what happens subsequently.

Through birth, Dasein finds itself in a particular generation with a particular destiny. Destiny ought not to be thought of deterministically. On the contrary, Heidegger understands fate and destiny in terms of freedom. In freedom, historicity and practice come together. Rudolph Brandner interprets Heidegger's account of the relation of freedom and destiny in terms of freedom and necessity. (Brandner, 1994, pp. 135–8) He sees the phenomenal basis of necessity in the facticity of Dasein. Necessity here is understood as what Aristotle called 'conditional necessity': that which happens has not happened out of necessity, but once it has happened cannot be undone. The point in time t(x) cannot be changed. Historicity, for Brandner, can be understood in these terms: 'In it [historicity] the becoming of human beings accomplishes itself to that which it is, as the constant transformation of possibilities into the irreversible facticity of its being-in-the-world.' (Brandner, 1994, p. 137) The human happens in its being through the transformation of possibility into irreversible facts. This, Brandner suggests, is what Heidegger means by fate and destiny. (Brandner, 1994, p. 138) According to this account, Heidegger understands historicity in terms of an Aristotelian account of possibility. If this were true, the historicity sections of *Being and Time* would mark a reversal in Heidegger's account from his previous transformation of the concept of possibility.

Heidegger addresses this question quite directly. He states: 'To expose the *structure of occurrence* and the existential and temporal conditions of its possibility means to gain an *ontological* understanding of *historicity*.' (BT, p. 344; H, p. 375 [emphasis in original]) Thus, only through an understanding of 'occurrence' and 'possibility' can we comprehend historicity. Undeniably there are possibilities which can be actualized and are thereafter unchangeable. What happened on the 30 July 1994 belongs to the past. But, strictly speaking, Dasein has no past, because it is characterized by the movement which constitutes its existence. While it is true that for Heidegger existence is grounded in the temporal ecstasis of the future, 'having-been arises from the future'. (BT, p. 299; H, p. 326) If the having-been was really a sequence of actualized possibilities, then there would be no reason why the 'having-been' of Dasein would not consist of the making present of the past. Dasein can, of course, make present events from its 'past'. But it cannot appropriate these events as its own, because through making present, there is only an image of the past which can ultimately belong to anybody.[38] Such a past does not correspond to the having-been of Dasein. The having-been can only be accomplished through repetition.[39] Facticity can only be understood on the basis of such having-been: 'the primary existential sense of facticity lies in having-been.' (BT, p. 301; H, p. 328) If, furthermore, we must think of historicity, for Heidegger, from the having-been, then 'occurrence' and 'possibility' can only be understood on the basis of the movement of the having-been, that is, from repetition, which is accomplished in the moment.[40]

If all of this is true, then we must think destiny and its relation to freedom differently than Brandner does. The necessity of which he speaks is that of inauthentic historicity, which sees the present as a mere actualization of what has been. Brandner sees freedom as the capacity to change necessity into possibility. On the contrary, facticity is itself already possibility, but not possibility in the sense of *potentia*.

The question remains, how we should understand destiny and fate. According to Heidegger, 'as fate, resoluteness is freedom to *give up* some definite resolution, as may be required in a possible situation.' (BT, p. 357; H, p. 391 [emphasis in original; translation modified]) Fate is the freedom towards a specific action which changes according to the situation. Although Heidegger does not explicitly speak of action here, this relation of freedom to fate can only be understood on the basis of a concept of action or practice. In describing the fateful character of Dasein, Heidegger states: 'Existing fatefully in resoluteness handing itself down, Dasein has been disclosed as being-in-the-world both for the "coming" of "fortunate" circumstances and for the cruelty of accidents.' (BT, 351; H, p. 384 [translation modified]) Heidegger here stresses a certain passivity, but also a readiness to handle cruel and good fortune. As fated, Dasein does not have the possibility to take a distance from its own situation; it must respond. As authentic, it responds through the repetition of possibilities of having-been Dasein. Fate is not simply that which happens, rather Dasein must choose its possibilities, to be fated. But this choice is viewed as practice. Heidegger states that Dasein chooses its 'heroes'. (cf. BT, p. 352; H, p. 385) By 'hero', he means that which breaks from the everyday, those great possibilities of the having-been, which are hidden through the levelling off of tradition. This does not mean a blind repetition of past 'great events' which should serve as examples for the present. Rather, Heidegger means that through an engagement with the past – Heidegger speaks of making a rejoinder (*Erwiderung*) to possibilities – the today as the moment would be freed up to leave open a space for practice.[41] This space is that of possibility. As Heidegger puts it: 'the handing down of a possibility that has been in retrieving it, however, does not disclose the Dasein that has been there in order to actualize it again.' (BT, p. 352; H, p. 385)

'Possibility' is nothing objectively present, nothing that has been 'actualized'. In the moment, Dasein comes to the decisive possibility, to be historical; that is, experiences in the moment the possibility of a change, which concerns the possibilities of existence of a community. This experience is that of the 'simplicity (*Einfachheit*)' of fate, which Heidegger emphasizes. Dasein is snatched back from the endless tasks of everyday life to its own finitude. Dasein is brought to a point of rupture, analogous to death. As with death, an abyss is opened up, namely that of the groundlessness of existence. In other words: the moment is a movement of occurrence in which Dasein, through repetition, comes into a new order.

This stress on the abysmal in the moment may seem to be in tension with the stretching constancy of the connectedness of life, which Heidegger attributes to the moment. Heidegger indeed stresses the unity of the connectedness, not its fragility. But it is precisely here that Heidegger is determining the difference between kairological and chronological temporality. The constancy of stretching does not mean constancy *in time*. The unity is not the unity of a 'series of "experiences" that has ensued and is still ensuing which can be subsequently linked together' (BT, p. 356; H, p. 390 [translation modified]). The connectedness in question is not one between previously disconnected events, but rather the unity of the moment. It is a unity of birth and death, understood as possibilities of finitude. This does not mean primarily that Dasein was in fact born and must die, but rather that, as possibility, birth and death are contingent, that it constitutes the contingency of existence. With respect to this contingency, birth and death both name the same thing, namely finitude.

In the everyday, Dasein is dispersed. Dispersion (*Zerstreuung*) is, for Heidegger, primarily a temporal term and refers to the chronological sequence of this or that actualized or non-actualized possibility. The unity that is possible here is only a subsequent unity, which is always subject to the danger of again being dispersed. To the extent to which the historian understands history as a sequence of facts, which only have a subsequent unity, she is held captive by such chronology.

The alternative here is not between unity and disarray, but rather between two different senses of unity. The abysmal to which I have referred above is not disorder, but rather an order without ground. This order is reflected in the kairos. That which has stable ground and order, namely the one-self (*das Man*), is seen from the point of view of authenticity to be merely a superficial order, and indeed a flight from death.[42] Dispersal assumes this order, because only in the order of the one-self can Dasein be dispersed in its tasks. Dasein confirms this order, in its subsequent establishment of a chronological connectedness, without being aware of the abysmal in the dispersal.[43]

The moment of vision is the 'place' of repetition. Heidegger describes repetition as follows:

> Repetition makes a rejoinder to the possibility of existence that has-been-there. But the rejoinder to the possibility in a resolution is at the same time, *as in the moment of vision, the disavowal* of what is working itself out today as the "past". (BT, pp. 352–3; H, p. 386) [emphasis in original; translation modified])

Repetition so understood assumes a past not of actualities, but of possibilities which address Dasein in the present. Authentic Dasein is free from the past as levelled off in tradition, when it attempts to begin anew, that is, when in the moment it grasps possibilities which have-been, in order to bring about something new. It is only because Dasein is historical and can overcome the old that it can recognize and bring about the novel, that is, that it can have temporality in the sense of kairos.

The choice of fate is the choice to be such possibility. The goal of such a choice is neither to narrate a history (*poiesis*) nor to engage in theoretical knowledge (*epistēme*), but rather to make practice historical. The understanding of practice in *Being and Time* is highly influenced by Aristotle, as we will see in the next chapter. But in understanding practice historically, he goes beyond Aristotle, not simply because he ontologizes the latter, as Volpi (cf. Volpi, 1988) contends, but again because he overcomes Aristotle's account of possibility as potentiality. It is characteristic of action that things can be otherwise, but only within the existing possibilities of action. By thinking action historically, Heidegger can understand it as change and movement without any essential ground.

In considering the lack of ground – the abyss (*Abgrund*) – we need to focus again on the difference between death and birth. Both birth and death have an abysmal character. To understand this in relation to practice, it his helpful to relate it to Arendt's account of birth as the capacity to begin. As is well known, Arendt returns to Augustine's distinction between the beginning which is the human (*initium*) and the beginning of the world (*principium*). (Arendt, 1998, pp. 177–8) This is a difference, according to Arendt, between the beginning of something and the beginning of someone – the human being as beginner, initiator. The beginning of someone is repeated in every human birth. A beginning is something new. The meaning of the new can perhaps best be understood as an interchange of production and reproduction. (cf. Waldenfels, 1990, pp. 95–6) We speak of birth as reproduction, but it is at once a new production. In repeating the past, something new and unprecedented emerges. Birth as such is not simply the point in time in which life begins, but is much more the possibility to begin, which is repeated in action.

'These fates have already been guided in beforehand in being-with-one-another in the same world and in the resoluteness for definite possibilities.' (BT, p. 352; H, p. 384 [translation modified]) Through birth, Dasein is in the same world with others, it makes such being-with possible. Coming into the world through birth, Dasein is in the world with others. To take up birth (*die Geburt* <u>*ein*</u>*holen*) means to take up again or repeat (<u>*wieder*</u>*holen*) the world of being-with. Understood literally, 'tradition' is the handing over (*tradere*) in birth of one generation to another. Nobody can in this sense hand themselves over, although Heidegger does speak in those terms. (In repetition, we hand ourselves over in a certain sense, but repetition assumes the passivity of birth as its prior condition.) Heritage is made possible through birth. The German word '*Erbe*' means both what is bequeathed (heritage) and the one who inherits (the heir).[44] Only if someone is born (the heir) can there be anything to hand on (heritage). In those cases where a culture has vanished and only ruins give witness to it, it has often been attempted to more or less artificially produce connections with that culture. Only then can the new culture establish itself as a repetition of the old. But the very artificiality of that attempt points to the fact that historical repetition is possible only through generativity, that is, through birth.[45]

Dasein is authentically historical only in the repetition of its own birth and in its beginning. The moment of repetition is fateful in the sense that it gives the chance to glimpse the hidden possibilities – metaphorically expressed, the *fortuna* – of the past. This implies an understanding of time as fulfilled rather than empty. As such we understand time as destiny. The root meaning of the German terms *Schicksal* and *Geschick* is, as we have seen, to be prepared to reach a certain goal or achieve a certain end. A destinal time is, as such, a time which is appropriate for something. It is also a time which removes Dasein from the chronological. It is the moment of at once beginning and ending, hope and despair. It is the moment in which Dasein is brought before radical change both in itself and in the world as experienced by its community.

II.2 Conscience and kairological temporality

Heidegger states repeatedly in the chapter on historicity that historicity is grounded on temporality.[46] However, his actual account of historicity seems to contradict this claim, as it brings into play elements which are not grounded in temporality as he describes it. The question now is whether Heidegger's account of temporality, read backwards from his account of historicity, can be interpreted as *historical time*.

I have stressed above that historicity needs to be understood in terms of a concept of action, but if that is the case then the aspect of authenticity becomes problematic. Heidegger's account of authenticity assumes that it is thinkable for an authentic existence to be in the world as an individuated Dasein. He states explicitly that this individuation which is grounded in being towards death reveals the fact that 'any being-together-with what is of concern and any being-with the others fails when one's ownmost potentiality-of-being is at stake.' (BT, p. 243; H, p. 263 [translation modified]) Heidegger has been frequently criticized for founding authenticity on being-towards-death. I am not going to repeat that criticism here. What is not so frequently remarked upon, however, is that historicity is a phenomenon of authentic being-with. Put briefly: in *Being and Time*, Dasein is both temporal and historical, but while the tendency of the temporality account is that Dasein finds its own potentiality-of-being in its own being responsible for and to itself, it lives its historicity only through acting co-responsibly with others in a common world. The question now is whether there is anything in the discussion of temporality which could account for such action.

This question can be responded to only with a firm view on the kairos as authentic temporality. The question then becomes, can the kairos be understood in such a manner which justifies the grounding of historicity on temporality. It is characteristic of kairos that time and action become one: to act in the kairos is to act in response to the opportunity of the moment; the opportune is that which is new in the moment. How then can we account for novelty?

Heidegger discusses novelty or the new in the context of curiosity (*Neu-gier*: greed for the new). In this account, it appears that the new is only to be understood as inauthentic. But it is Dasein's way of being toward the new which is here at issue. As Heidegger puts it: 'It [curiosity] seeks novelty only to leap from it anew to another novelty.' (BT, p. 161; H, p. 172 [translation modified]) Heidegger does not tell us directly what an authentic relation to the new would be. But in both the sections on conscience and on falling (*Verfall*) he refers to the new and hints towards an authentic 'being-towards-the-new'.

In the sections on everyday temporality, Heidegger seeks to show how it is derived from original temporality. In § 68, he discusses inter alia the temporality of falling. Falling has the existential sense of the present, and in his analysis Heidegger concentrates on curiosity as a mode of falling. He discusses curiosity on the basis of seeing. He stresses, however, that 'seeing' here is not perceiving through bodily senses, but rather 'perceiving in the broader sense lets what is at hand and objectively present be "bodily" encountered in themselves with regard to their outward appearance.' (BT, p. 318; H, p. 346) In seeing, and specifically in curiosity, Dasein encounters the at-hand and the objectively present; specifically it encounters them in their novelty. But such a seeing allows the thing to be encountered not in order to understand the new, but only 'in order to see and to have seen [it]'. (BT, p. 318; H, p. 346) Furthermore, although curiosity is orientated towards the future, it is so inauthentically because it seeks to make the future present, and as such does not wait on possibility. In curiosity, the future is simply a not-yet-occurred.

The new as such is not inauthentic; the inauthenticity of curiosity consists in the fact that curious Dasein does not recognize the new as such. Curiosity does not dwell with the new but rather immediately goes on to the next. This shows that in curiosity Dasein does not give itself to the thing itself (*Sache*), in this case the new. Dasein is fallen in the new, because it does not tarry with it. As dispersed and not dwelling, it is without abode. The present in this mode is the extreme opposite to the moment of vision. This is so because Dasein does not hold fast to the present. To do so would be to be ready to abide with the new, to wait on it, so as to be able to act in the moment.

In the context of this analysis, it is important to note that Heidegger represents conscience as practically the opposite way of being towards the new. In Part 1 of *Being and Time*, he discusses idle talk (*Gerede*) and curiosity as two forms of falling; the one associated with hearing, the other with seeing. While curiosity sees the new, idle talk hears it. In idle talk, Dasein fails to hear itself. Conscience is that which calls Dasein out of this being lost in the 'everyday new' of idle talk. (BT, p. 251; H, p. 271) The one called is Dasein; the call comes, however, not from outside, but from Dasein itself. The called self is called to itself and that means to its own potentiality-of-being. It is Dasein itself which calls, but Dasein is for the most part caught up in the one-self: the call of conscience calls Dasein out of its absorption in the one-self.

Conscience discloses itself as call. The call is a mode of discourse.[47] The disclosure in question here is possible only because Dasein is 'in the truth', that is, Dasein is an entity for which being can be disclosed. As is well-known, Heidegger argued that primordially truth is not propositional, but rather should be understood as unconcealment (ἀλήθεια). As such discourse is not the saying of sentences and propositions, but rather the articulation of that which happens in the opening, in which the possibility of true and false sentences are opened up. (cf. BT, pp. 150–1; H, p. 161) Dasein hears in conscience a discourse, which opens up truth in this primordial sense. Before discussing the sections on conscience in *Being and Time* in more detail, we can take a point of orientation from a lecture course on Plato's *Sophist*, which Heidegger held in the winter semester of 1924–5. In that course he shows how, in his discussion of φρόνησις, Aristotle came upon the phenomenon of conscience. (cf. McNeill, 1999, pp. 114–5; Bernasconi, 1989, pp. 127–47) Decisive in Aristotle's account of φρόνησις, according to Heidegger, is the impossibility of forgetting. While, for example, a producer can forget that which he knew about a particular product, what a person knows about practice cannot be forgotten, because it is a knowledge concerning herself.[48] This is the case because, in a certain sense, there is nothing to forget. Φρόνησις is not a register of objective knowledge, which may be forgotten. Rather, φρόνησις is the ability to consider and to deal with changes. It is the ability to reflect on action. Such reflective thinking seeks to understand how one can and should act. That has to do with the kairos, to the extent to which φρόνησις is the ability to deal with change, that is, to cope with the new. (cf. Aubenque, 1976, pp. 95–106) Φρόνησις as such is the virtue of finitude. As finite, the human being has an uncertain relation to the future. *Φρόνησις* is the ability to see the possibilities of a certain time, and the knowledge of how one can and should act in that situation. (Heidegger, 2003, p. 39; GA 19, p. 56) The inability to forget is the capacity to act in the moment in relation to Dasein's own preparedness for the situation. *Φρόνησις* here reflects the call of conscience, which does not allow for a real forgetting, a forgetting which forgets itself.

Both the identification of conscience and *φρόνησις*, and the stressing of the impossibility of forgetting, seem at first sight to contradict the line from Goethe which Heidegger constantly alludes to in this context: 'he who acts is always without conscience (*der Handelnde ist immer gewissenslos*)'.[49] *Φρόνησις*, however, is not action itself, but thinking concerning action. Such thinking is always subject to uncertainty. It cannot ground action. It is a thinking of contingency, uncertainty and the unforeseen. In order to be able to act, Dasein must possess the possibility to forget, that, for example, its action may harm another in its Dasein. (cf. Heidegger, 1985, p. 319; GA 20, p. 441). This means that it must let itself be involved in a situation, let itself be carried away by the situation, without heed to the future, chronologically understood.[50]

On conscience, Heidegger says: 'the call comes *from* me, and yet *over* me.' (BT, p. 254; H, p. 275). This is kairologically structured. The call happens in the moment. If we understand conscience in this respect, then the call of conscience is uncanny (*unheimlich*) because it brings Dasein before the new or the force of change. 'The caller is unfamiliar to everyday one-self – it is something like an *alien* voice.' And he goes on: '"It" calls, even though it gives the concernfully curious ear nothing to hear which might be passed along and publically spoken about.' (BT, p. 255; H, p. 277) It is not by accident that Heidegger again here distinguishes curiosity from conscience. Conscience is a form of discourse (*Rede*), curiosity a form of idle talk (*Gerede*). Fundamental here is the place of hearing: hearing for the alien, for that which breaks with the normal, the new, which is not heard, but quickly passed over in curiosity.

In the wake of the discussion of conscience, at the beginning of chapter 3 of the second division of *Being and Time*, Heidegger poses a remarkable question: 'What is death supposed to have in common with the "concrete situation" of acting?' (BT, p. 279; H, p. 302) This question only makes sense if conscience and action are essentially connected. In discussing the totality of Dasein, Heidegger brings into a relation of co-originality, guilt, conscience and death. Conscience is thinking on death. In being guilty, Dasein encounters the nullity (*Nichtigkeit*) of freedom. 'Freedom *is* only in the choice of the one [possibility], that is, in bearing the fact of not having chosen and not being able also to choose the others.' (BT, p. 263; H, p. 285) Freedom is, in that sense, null. Freedom shows that this nullity comes into Dasein not only through death, but also with the call, which calls Dasein to its own guilt and, as such, to its freedom. This nullity becomes clear when Heidegger interprets Dasein's 'becoming free for the call' as its 'readiness for the potentiality-of-being-called ... It [thus] has chosen itself'. (BT, p. 265; H, p. 287 [translation modified]) This means that Dasein experiences its freedom and finitude in its preparedness for the moment of the call. The understanding of the call is itself the choice of having a conscience. Understanding the call means wishing to have a conscience. This Heidegger understands as 'readiness to be called'. (BT, p. 265; H, p. 288)

Decisive here is that discourse and understanding both require a readiness to listen. This readiness is an openness to hear the call. Heidegger hints at how it is possible to have such readiness when he says, 'to hear the call authentically means to bring oneself to factical action.' (BT, p. 271; H, p. 294) Hence, the readiness to hear the call is the readiness to act. That does not mean that the call of conscience gives Dasein practical instructions as to how to act. Rather, wishing-to-have-a-conscience means that Dasein wishes to be ready to be attentive to the time, in which the moment arises, which is appropriate for certain actions. What is uncanny in conscience is that it shows the human being that time cannot be reckoned with, and that Dasein should always prepare itself to care for the new.

This becomes clear in the next sections concerning resoluteness (*Entschlossenheit*). He says, 'What we call accidents (*Zufälle*) in the with-world and the surrounding world (*Mit- und Umwelt*) can only *be-fall* (*zu-fallen*) resoluteness.' (BT, p. 276; H, p. 300) As we have already seen, such resoluteness is fundamental to fate. Here it is clear that not only can the historicity of Dasein be understood kairologically, but also that authentic temporality can be defined in terms of a recognition of the kairos. Furthermore, as the word 'accident' (*Zufall*) suggests, Heidegger is attentive to the ambiguity of the kairos as a time both of opportunity and of misfortune.

The kairos demands readiness for that which we cannot predict, a true readiness which does not simply attempt to make the future present in the mode of its past. Resoluteness for Heidegger is essentially defined by such readiness. He defines resoluteness as Dasein's 'reticent projecting itself upon its ownmost being-guilty, in readiness for Angst'. (BT, p. 273; H, pp. 296–7 [translation modified]). Understanding the call discloses to authentic Dasein its own uncanniness. That disclosure is a finding of itself of Dasein in a situation, through a particular affective state (*Befindlichkeit*).[51] The readiness to be addressed is at once a readiness for Angst, and as such Angst is the mood of conscience. Heidegger speaks even of the Angst of conscience (*Gewissens*angst). As such wishing to have a conscience is a readiness for Angst. But whereas the sections on Angst in *Being and Time* stress the individuating force of that mood, when we read this readiness for Angst in the context of resoluteness and conscience a different picture emerges. According to Heidegger, 'it is from the authentic being a self of resoluteness that authentic being-with-one-another first arises'. (BT, p. 274; H, p. 298) Such authentic being-with Heidegger understands as solicitude (*Fürsorge*). Authentic solicitude lets the other be itself. It gives the other Dasein back its care. This can only occur within a shared situation. In resoluteness, Dasein is already placed in a situation. 'Situation' here does not mean simply position in any formal sense, but rather a specific place, hence a specific spatiality in which Dasein encounters an other. Conscience calls Dasein into such a situation. That means that Dasein is placed in a situation through conscience. As such, resolute Dasein is not simply in a situation, but acts already in the situation. The situation is revealed through resoluteness, in which Dasein responds to the being addressed in conscience. The situation is revealed only in action.

Heidegger expressly stresses that, in these very passages of *Being and Time*, he is avoiding the term 'action'. (cf. BT, p. 276; H, p. 300) He does this to avoid the misinterpretation that he is simply making a pragmatic claim about the priority of practice. He is making, rather, a claim about the finite temporality of Dasein. Such a finitude is prior to any distinction between theory and practice. The difficulty is to think practice before this distinction, to think, in other words, its pre-structure. Because of this difficulty, practice remains 'unsaid' in *Being and Time*. But for all that, this work does not just contain an implicit understanding of practice, but also a concept of the truth of practice, or better, practice as the place of truth. To grasp this, we need to return again to the phenomenon of birth.

I have already pointed out that birth is lacking in the analysis of temporality. To try nonetheless to understand birth within temporality in Heidegger, we need to turn again to Arendt. Heidegger says: 'Factical Dasein exists natally and natally too it is already dying in the sense of being towards death'. (BT, p. 343; H, p. 374 [translation modified]) Arendt, in apparent contradiction to this, says: 'the human being must certainly die, but he is not born to die, but rather on the contrary to begin with something new.' (Arendt, 1998, p. 246) According to Arendt, human birth opens up the possibility of acting in the sense of being able to begin. It is the capacity to interrupt the biological cyclicality of life. Only as a natal being is it possible to begin, that is to bring about something new.

The two statements of Heidegger and Arendt are not irreconcilable. Heidegger's claim is that to be natal is to be finite; Arendt's claim is that to be human is to be able to act and to begin. For Heidegger, the finitude of Dasein means that it is conditioned by the possibility of its own capacity to be. But if we understand such possibility not from the standpoint of the possibility of impossibility, but from that of the possibility to begin, then we can understand the finitude and temporality of Dasein as natal and not simply mortal. If this is so, then action plays an essential role in the finitude of Dasein.

We are now in a position to understand Heidegger's question concerning the commonality between death and action. Heidegger himself reformulates this question as one of the connectedness between anticipation (*Vorlaufen*) and resoluteness. (cf. BT, pp. 279–80; H, p. 302) If resoluteness is a readiness for finitude, then the question here is a question concerning the temporality of Dasein, as an entity which encounters its finitude in action. Such a temporality is that of the kairos, which resoluteness first discloses. Resoluteness is not so much Angst as *readiness* for Angst. This readiness can be understood in connection with the new, that is, with that which Dasein encounters in beginning. To see this, we need to interpret resoluteness in relation to birth. In doing so, I am understanding the being towards end as a turning *back* towards birth and not simply as an anticipation of death.

Resoluteness is Dasein's 'letting itself be called forth to its ownmost *being-guilty*.' (BT, p. 283; H, p. 305 [translation modified]). Understood in relation to birth, this would mean that Dasein is disclosed as guilty when it begins or is ready to begin. This interpretation seems less arbitrary when we see how closely Heidegger relates being-towards-death or being-towards-end with the call of conscience:

> Understanding the call of conscience reveals the lostness in the one-self. Resoluteness brings Dasein back to its ownmost potentiality-of-being-a-self. Its own potentiality-of-being becomes authentic and transparent in the understanding being-towards-death as the *ownmost* possibility. (BT, p. 283; H, p. 307 [translation modified]).

Resoluteness is pivotal here. The call of conscience calls Dasein out of its lostness in the one-self. But resoluteness is necessary in order to bring Dasein back to its own capacity to be. In resoluteness, Dasein chooses to have a conscience. That choice is the choice for its own being guilty. Understood in temporal terms, this is the readiness to act in relation to an uncertain future in which Dasein must bear the guilt of its actions. Such action is guilty insofar as in beginning it brings about that which it cannot know. But the action is not arbitrary; it rather responds to the call of conscience.

In *The Basic Problems of Phenomenology*, Heidegger states that resoluteness has its own temporality, which shows itself in Dasein's coming back to itself. The future of such resoluteness he characterizes as 'anticipation', the having-beenness as 'repetition'. The present he characterizes as the moment, and identifies the moment of vision as kairos. The moment of vision is 'that which arising from resoluteness, has an eye first of all and solely for what constitutes the situation of action.' (Heidegger, 1988, p. 287; GA 24, p. 407 [translation modified]). He goes on: 'The moment of vision is an original phenomenon of original temporality ... Aristotle has already seen the phenomenon of the moment as kairos.' (Heidegger, 1988, p. 288; GA 24, p. 409 [translation modified]) If resoluteness determines action, then the moment is constituted as natality, because the situation is always new and other. The now, on the other hand, is always the same and can be calculated. A situation is, in its own factical possibilities, never the same as another. As resolute, Dasein takes up the possibility of death in its own capacity to be. As historical, Dasein takes up its birth in the same manner. Dasein does this in the resolute moment. Only in this moment, through the resolute response to the call of conscience in a situation, does Dasein encounter its own finitude.

'With the phenomenon of resoluteness we were led to the primordial *truth* of existence.' (BT, p. 284; H, p. 307) This truth is the temporality of Dasein which defies all reckoning, disclosing the meaning of authentic care. Truth is revelation and Dasein is at once the revelation and the being revealed of its own capacity to be. Heidegger does not mean that there is truth which is indifferent to situation, but rather on the contrary that truth is essentially tied to the specific situatedness of Dasein, and hence 'resolution must be *held open* for the actual factical possibility in accordance with its own meaning as a disclosure (*Erschliessungssinn*).' (BT, p. 284; H, p. 307 [translation modified]) This means the disclosure of the situation of action is the truth – as unconcealment – of practice.

The temporality of Dasein is only authentically temporalized to the extent that Dasein is individuated. Being-towards-death is only experiencable for individuated Dasein. 'Individuation does not clinging obstinately to one's own private wishes but being free for the factical possibilities of current existence.' (Heidegger, 1988, p. 288; GA 24, p. 408) In *Being and Time*, it is not clear how in such freedom Dasein can place itself in being-with-others; how, in other words, it can ever transcend its own private wishes, no matter how they manifest themselves.

Indeed, in his discussion of historicity, Heidegger hints at this difficulty: 'Destiny is not composed of individual fates, nor can being-with-one-another be conceived as the mutual occurring of several subjects.' (BT, p. 352; H, p. 384 [translation modified]) At the end of this sentence, a footnote refers the reader back to § 26, in which the phenomenon of being-with is made thematic for the first time in *Being and Time*. It seems to me, however, that an important shift in Heidegger's understanding of being-with occurs between these two passages. In § 26, being-with is not an 'occurring together of several subjects' because Dasein is original being-with, and therefore being with another is not dependent on the coming together in a group of several subjects. But Dasein is historical because it is destinal. And it is only destinal because it is with others in the same world. Hence, being-with as understood historically is not an occurring together of several subjects, because Dasein is *always already* a member of a generation *through its birth*, which precedes it and which precedes any multiplicity of 'subjects'.

As we have seen, central to this is the ambiguous concept of heritage/heir (*Erbe*). Heritage is constituted by the 'goods' which are handed down, and the character of these goods lies in the 'making possible of existence'. The goods, which are the heritage, constitute the heir and do so in a fundamental ontological sense. The making possible of authentic existence constitutes itself in resoluteness and is, therefore, the handing over of a heritage. Heidegger insists that such resoluteness be understood in the context of being towards death. As such, he sees 'the primordial occurrence [*Geschehen*] of Dasein' as the occurrence in which Dasein '*hands itself down* to itself, free for death, in a possibility that it inherited and yet has chosen.' (BT, p. 351; H, p. 384) Being free-for-death is made possible through an inherited possibility, through Dasein's heritage. This is possible only for Dasein as belonging to a generation. Through birth, Dasein is precisely not individuated, but is bound to others. To take over its birth is for Dasein to take over, or rather 'repeat', the world which it shares in being with others.[52]

In contradiction to Heidegger's claims, then, far from historicity being grounded in temporality, temporality seems to be grounded in historicity. But perhaps the very question here is mistaken: might it not be that Dasein is ontologically and originally both temporal and historical? Dasein would then be historical on account of its birth into a generation, and would through that (following Arendt's account of the capacity to begin) have kairological possibilities. Conversely, if *Dasein* was not itself temporal, it would have no relation to history, but would simply be a being in time without access to the past and would as such be – like Nietzsche's cows – fully dominated by forgetfulness.

III The order of time and the historicity of Dasein: The transcendental and authenticity

My purpose in this chapter was to treat *Being and Time* as questionable, or as German more elegantly puts it, 'worthy of question (*fragwürdig*)'. The questions which I have posed will concern us for the rest of this book, and also when we begin to go beyond *Being and Time*.

The question of the order of *Being and Time* is one which hints at a tension underlying the work itself. What we sense in that work is what Heinrich Hüni revealingly terms a 'desire for structure (*Baulust*)'. (Hüni, 1985, p. 35) The work is systematic and is constructed along a certain blueprint. It is – if the analogy is allowed – more an American city like Manhattan or Chicago than a city from the old world such as Rome or Jerusalem. But somehow the architecture fails. The city of *Being and Time* remains incomplete. It is perhaps not an exaggerated claim to say that this incompleteness has to do with a failed correspondence between the order of the book and that of the things themselves. The attempt to found historicity on temporality is consistent with the blueprint of *Being and Time*, but is a forced ordering, which fails to fully take historicity seriously.

The order of time, which resists the ordering tendency of *Being and Time*, is that of a split order of chronos and kairos. In both, there is an interplay of unity and difference. The moment of vision understood as a leap puts the chronological order into question and can be understood only out of that difference with chronology. It has in itself no united ground. It is not quantitatively extended, even when years go by, as we hesitate in the kairos.[53] The kairos is rather qualitatively differentiated; its unity consists in this differentiation. Simmel's essay on adventure can again illustrate the phenomenon in question. According to Simmel, the adventure has a unity similar to that of an artwork. As Simmel describes this analogy, there is a hint of the relation of kairos to chronos. (cf. Simmel, 1997, pp. 13–14) According to Simmel, the essence of the artwork lies in the manner in which it cuts a piece (*Stück*) from the continuity of lived experience. This piece is removed from its connectedness with past and future and takes on a certain form, and it is this perception of a part of existence as forming a closed unity which the artwork and adventure have in common. In other words, the artwork and the adventure share the structure of kairological time. They are not outside of time, but rather outside of time chronologically understood. They have in common a leap, which forms a break with past and future, but in that very break relate to both in the manner of an accentuation. This accentuation is at once becoming alien, and it is the kairological in both art and adventure which is found in their manner of making every relation – above all temporal relations – strange and alien. When Heidegger speaks of such experiences, he does so in terms of mood. The next chapter will outline how Heidegger's discussion of mood shows the kairological manner in which, through an alienation of the past order and revealing the full indeterminacy of the future, a new unity and a new connectedness of past and future is formed.

But the kairological cannot be understood except in relation to the chronological from which it marks a rupture, a break and a new beginning, which again is subject subsequently to the retrospective force of chronology. Nonetheless, it is only because Dasein can experience the abysmal in history that it first has history. This is rooted in birth itself, the entrance through generativity into history, which is characterized by the abysmal. Here again, we are faced with the problem of order. Just as with the relation of temporality and historicity, the relation of kairos and chronos is not a foundational one, but rather a crossing over of one into the other, an interplay. If that is so, then the question cannot be avoided as to whether the relation of ontic and ontological needs to be understood in this manner also; whether the structure of *Being and Time* is too vertical, and not sufficiently conceived horizontally. One may further ask if this vertical construction of the work is not in a certain tension with the circularity of understanding, which in many ways guides the analyses of Heidegger's first magnum opus.

Fundamental here are two issues, which are deeply related: the transcendental structure of *Being and Time* and the meaning and significance of authenticity. These belong together; the difficulty is to work out how.

Being and Time is a transcendental work in a very specific sense, a sense which is tied up with the question of authenticity. Although it clearly has normative content, authenticity is not a *morally* normative concept. To be authentic is not to be morally superior; it refers not to the human person as such in her moral, political or social life, but rather to the Dasein in the human being. That which is authentic to Dasein is its own; its own is the question of being. Authentic Dasein is that Dasein which questions the 'there' of its being, which thinks transcendentally, which thinks not about what is actual, but rather about the possible as possible. The normativity here is a philosophical one: to be authentic is the precondition for thinking philosophically. Steven Crowell puts this well: '"Authenticity", a clear view of one's own being, is a condition of philosophy's possibility.' (Crowell, 2001, p. 210) Philosophical thought is only possible for a Dasein which understands its own situatedness. The transcendental conditions of Dasein's being are those of a temporal and historical being. As such, authentic thought is thought of situatedness. The authentic being in that situatedness is one which sees itself as situated, and which sees its situation in terms of that insight. The transcendental is not a theory here, but rather the conditioning possibility of both the appearance of things for Dasein *and* its action on things. Authenticity is a modification of inauthenticity, and can only be understood in the tension of that relation. But nothing of content changes between authenticity and inauthenticity: the authentic is a view of the world which sees it in terms of Dasein's own responsibility for it. This responsibility is not simply for itself, but rather also for the past possibilities, the historical possibilities, which it inherits. In engaging with its situation, Dasein engages its own being as natal, its own transcendental structure – its being toward possibility as possibility – in the everyday world. In doing so, it acts in and out of possibility and distances itself from the fetishizing of the object. But it does this not out of a disengagement with things, but rather in its realization of the thing's transcendental structure – that meaning as possibility is generated out of the thing as it appears to Dasein in its acts of transcendence.

As such, authentic temporality is the temporality of the making possible of the question of being. That which makes possible the question of being is the rupture in the everyday, a rupture which is only thinkable in temporal terms. But this rupture, while allowing Dasein to gather its temporal being from dispersal in things, does not mean a removal from engagement with things, but rather opens Dasein up to authentic action. The encounter with being is an encounter in action, which in this sense is a grasping of an opportunity. As authentic, Dasein encounters the situation of its existence as a possibility. In this sense, the opportune is constituted by a qualitative break in time. Clearly what is meant here is not the opportune in the sense of that which is thrown up by the actualities of the world of Dasein's concerns.[54] The claim is rather that the opportune moment is one of an ontological event, the occurrence of a historical, indeed epochal, transformation of meaning. The opportunity for philosophical action lies in responding to such transformations through a rejoinder of the possibilities of the having-been.

The problem here is one of how to understand the foundational relations of *Being and Time*: authentic temporality is that temporality which structures the being of Dasein in opening itself to the ontological question of the meaning of being. At that level Dasein does act on entities, on things, but does so in relation to the possibility in which the contingency of their actualities is revealed. The transcendental mixes with the empirical, the constituting mixes with the constituted in the historicity of Dasein, which is both transcendental and contingent.

The intersection of the transcendental and the authentic – precisely because authenticity is not a moral or political term – relates back to the moral and the political in ways which we will explore in the course of this book.[55] Authentic Dasein is that Dasein which suspends its daily concerns, suspends questions also as to how it ought to act within the constraints of those concerns, in order to ask the question of being. That rupture of commitments allows Dasein to free itself from its time precisely when that time is in crisis. The crisis becomes philosophical to the extent to which it concerns the very destiny of humanity itself. In that sense, Heidegger makes the transcendental historical without relativizing it; he asks what is it to be historical; what, in other words, does it mean to be in the world, that is in relation to being as contingent.

The tranquillity of the Kantian 'I think' is not available for Heidegger because he thinks Dasein in its continual fleeing from itself and as brought back to itself only in Angst. The moment of vision is itself a moment of Angst. But again this does not place Dasein beyond the world, but rather precisely in the world with the full weight of responsibility for the world. The kairos is the moment of this realization and the taking up of its ontological and historical task.

If this is the case, then the ontic and the ontological cannot be understood simply in a one-way foundational relationship. As Heidegger famously states: 'But does not a definite ontic interpretation of authentic existence, a factical ideal of Dasein, underlie our ontological interpretation of the existence of Dasein? Indeed.' (BT, p. 286, H, p. 310). This Heidegger understands to be a 'positive necessity'. Philosophy cannot escape its ontic situation, but must think from it, specifically from a historically conditioned ideal of authentic existence. But such an ideal arises out of Dasein's self-interpretation. In interpreting itself, Dasein understands itself from its own self-projection. Such self-projection is only possible because Dasein is in its transcendence, i.e. in over-stepping entities towards the being of entities. As such the ontological structures of Dasein are already implicit at the ontic level; at an ontic level, Dasein already glimpses, if obscurely, the transcendental structures of its being. The difficulty, however, as we will chart in the next chapters, is how to understand this transcendental structure of Dasein's being as radically historical, that is, as open to transformation in an event of appearance in which the world of entities transforms itself.

2

Praxis and Poiesis

The interconnection of kairos and chronos is not simply to be understood with respect to the relationship between temporality and historicity. This is so, above all, because the mode of experience in which the kairos comes to appearance has not yet been explored. It has already been provisionally stated that the basis for this experience is contained within practice. But the notion of practice remains undetermined so long as its difference from other forms of experience has not been made clear. Furthermore, practice can mean 'doing' (*πράχις*) as well as 'making' (ποίησις) in the sense of producing.[1] We may ask whether this latter difference between doing and making is significant with respect to the distinction between chronology and kairology. The thesis defended in this chapter is that this is in fact the case. In a very provisional and rough manner, my thesis can be formulated as follows: in the context of *Being and Time*, 'doing' is kairologically and 'making' is chronologically conditioned.[2]

Chronology within poiesis is characterized by the relation of poiesis to the past, and that the poietical order discloses itself as past, and indeed eternal. Furthermore, it will be shown that the poietical world is a lonely one in which there is no possibility for the radically new. Praxis, on the other hand, assumes a public world of being with one another and has a futural sense and directionality. In praxis, Dasein is not thrown or called back to an already past order. It is rather the case that the new order emerges in praxis out of the ruins of an existing order.

Praxis and poiesis are forms of Dasein's relation to the world. Dasein is not something which accidentally has a relation to the world. Rather, Dasein is 'being-in-the-world' (*In-der-Welt-sein*). That means that Dasein is only in its relations. Dasein is the *Da* of *Sein*, the 'there' of 'being', the 'place' in which the question of being first emerges. As such, Dasein is only in relation to being. Nevertheless, Dasein is concerned most immediately not with being, but with entities (*Seiende*). Entities are only within particular orders, which at a most basic level are temporal and spatial. As such, every entity refers beyond itself. This reference beyond itself does not simply refer to the horizon in which the entity is, but also to the mode of access to it which Dasein has. Dasein is only in terms of such access, being is only in the disclosure of entities. For this reason, Heidegger states that the question of the 'from whence' of the view towards being must be posed before the question concerning being itself. (cf. Heidegger, 1990, p. 153; KPM, p. 224) Being only discloses itself to human beings when the latter is truly Dasein, that is, 'in the truth'. Dasein is in the truth only in relation to entities, but then how is that relation to be characterized? This chapter will attempt to respond to that question. It is a question which ultimately concerns the temporal order of Dasein's relation to the world and to being.

This question is one which concerns Dasein's historicity, specifically the historicity of the question of being. All Heidegger's phenomenological descriptions of Daesin's relations to the world in *Being and Time* ought to be understood in the context of his programme of repeating and destructuring (*Destruktion*) the history of ontology. To support this contention, one need only look at Heidegger's 'translations' in that work.[3] One example of such a translation is the concept of 'handiness' (*Zuhandenheit*). In the introduction to his lecture course on Plato's *Sophist*, Heidegger described *τεχνή* as the self-recognition in concern and production. Production, he went on, was for the Greeks poiesis. With similar words, he described in *Being and Time* the relation with the ready to hand. (cf. Heidegger, 1997a, p. 28; GA 19, p. 40; BT, p. 63; H, p. 67)

I Handiness, objective presence and praxis

The relation of theory and practice in *Being and Time* is one which has excited much critical commentary. What is generally not recognized in this debate is that the theory/practice difference in that work is rooted in a more fundamental difference within practice itself between praxis and poiesis, doing and making. Understood in temporal terms, this difference is between the (poietical) priority of the past and the (praxical) priority of the future. The difference between practice and theory as, for example, it is understood by Gerold Prauss (cf. Prauss, 1999), is in *Being and Time* a difference in the realm of poiesis, which has its roots in Dasein's relation to eternity. In *Being and Time*, in short, the traditional difference of practice and theory is founded on what is presented as a more fundamental difference between praxis and poiesis. As such, the question is not whether theory or practice has priority in *Being and Time*, but rather what kind of theory or sight [*Sicht*] should guide Dasein in its authentic understanding – the praxical or the poietical? This question is important with respect to the understanding of time because, as will become clear, praxis and poiseis are structured differently in temporal terms, and furthermore the relation between kairos and chronos is only to be explained in relation to this difference.

I.1 Theory and practice

In his influential book, *Knowing and Doing in Heidegger's 'Being and Time'*, Gerold Prauss defended the thesis that Heidegger was seeking both to show a priority of practice over theory and then to dissolve this very difference. (cf. Prauss, 1999, p. 1) Prauss seeks to demonstrate that for Heidegger a reciprocal primacy exists between knowledge (*Erkennen*) and action. His analysis begins with an identification of dealings (*Umgang*) with action and circumspection (*Umsicht*) with knowledge as two interrelated domains. (cf. Prauss, 1999, pp. 2–4) This identification is already questionable, as circumspection is only then distinct from dealings when there is a rupture in action. At that point, circumspection crashes into emptiness. (cf. BT, p. 70; H, p. 75) [4] Only then does circumspection see 'what the missing thing was at hand *for* and at hand *with*'. (cf. BT, p. 70; H, p. 75) This clearly does not mean that until this point circumspection was blind, but rather that circumspection is now no longer the sight of dealings, seeing the 'for what' and the 'with what' as such – not in the purposefulness and directionality of action. In this sense, the theoretical aspect of dealing with the useful thing (*Zeug*), theoretical sight, is only subsequently recognizable. In the course of action, no distinction between 'circumspection' and 'dealings' is possible. As such, one cannot, as Prauss does, conclude that, on the basis of the fact that the dealings are not blind, there is a primacy here of circumspection over dealings. (cf. Prauss, 1999, p. 17)[5] Furthermore, it is apparent that Prauss identifies praxis with handiness. According to him, action is for Heidegger the actualization of a possibility. If, however, Heidegger describes Dasein as an acting being and understands the possibility of that being precisely not in terms of that which can be actualized, then one must doubt if Prauss has reached an original mode of engagement of Dasein at all. That which he interprets as the reciprocal primacy of knowledge and action is more fundamentally rooted in the transition from handiness to objective presence. If Dasein's being is not that which can be understood either as handiness or as objective presence, then one must ask whether knowledge and action cannot be understood more fundamentally than in terms of the difference and relation of theory and practice.

Carl Friedrich Gethmann understands the philosophy pursued in *Being and Time* as 'the earliest conception of a consistent pragmatism in the realm of

German thought'. (Gethmann, 1989, p. 143) At the same time, Gethmann makes clear that Heidegger did not think in terms of a distinction between theory and practice because he 'had a vulgar-platonic interpretation of this pair of concepts in mind'. (Gethmann, 1989, p. 147) Gethmann argues, however, that Heidegger nevertheless defended a methodological primacy of action, a primacy which could be of significance for both epistemology on the one hand, and politics and ethics on the other. Gethmann does not, however, pose the question whether action can mean the same in both domains – epistemology and politics/ethics. He understands action in Heidegger in terms of the 'central concept' of dealings (*Umgang*), which though needs to be understood in the realm of handiness. Is action with a useful thing and action with respect to an other the same? Are 'theories' which emerge in one or other domain to be evaluated in the same way? With these questions in mind, I will turn to the relation of handiness (*Zuhandenheit*) and objective presence (*Vorhandenheit*).

I.2 Handiness and objective presence

Time in the philosophical tradition is repeatedly subordinated to eternity. This has its roots in the ancient concept of theory: theory as a way of being, specifically of a way of lingering contemplatively with an object in a manner grounded in seeing. In order to reach such a state of lingering, it is necessary to take a distance from the temporal flow. Such a distance is only then possible if the object with which we delay is unchanging. If the object is unchanging not simply for a certain duration, but due to its own essential nature, then it was always there as identical. This must be because something which comes into existence is changed already by the very event of coming into existence. The object of theoretical contemplation, on the other hand, is simply present. Its presence refers to a specific relation to the past, namely that it always *was*, but never *passed*. In every point of time in the past it was potentially present for every observer. But how is *this* relationship to the past – to that which was and yet does not pass – possible when it does not correspond to human existence? In other words, in what relation to the world are we directed towards an original past, a 'past' which never passes? A recurring response to this question from Plato onwards is that this relation is one of recollection or unforgetting (ἀνάμνησις) (cf. Plato, *Phaedo*, 73c–74d), which is a mode of relating to the past which is not subject to temporal passing – not subject, that is, to the conditions of forgetting. Such unforgetting is possible only through the withdrawal from the temporal flow in theoretical contemplation. But such an account tells us nothing about the genesis of the object of our contemplative observation, or rather can speak of such a genesis only in mythical terms, i.e. in terms of a pre-birth which is prior to all possible experience. Although it may seem paradoxical to speak of the genesis of an eternal object, to speak philosophically at all of an object is first to secure the mode of experience of that object. Otherwise, the mode of being and temporal nature of the object is simply assumed rather than shown. Such theoretical contemplation remains with the presence of the object, and can understand its relation to the past of the object only through a projection of a past which never *was*, the past prior to all forgetting, that is prior to all temporal relations.

The first key step of Heidegger's analysis in *Being and Time* consists in rethinking the question of the genesis of the object in terms of handiness. It is here then that we must first seek the genesis of the eternal object for Dasein. Dasein encounters things not in isolation as objects of sense perception, but as things with particular purposes – things 'in-order-to' (*um-zu*). This is the mode in which the maker encounters things: her knowledge is not a theoretical knowledge, she knows only that a particular thing is useful in that it is suitable for a certain purpose. The eye of the maker is on this purpose or goal; as such the more inconspicuous the object, the more genuinely it is a useful thing. This quality of inconspicuousness is that which characterizes handiness.[6] The context of use is only intelligible in relation to the context of making. What is used is a product (a work) and this is used in order to produce another product. *Technē* is the mode of knowledge of such relations. It is the knowledge of poiesis, of the bringing forth into being, making. The useful thing is a work, but at the same time a means to produce other works. The work refers to the material out of which it was produced, and to the purpose for which it was produced. The work as such refers not only to the immediate environment (*Umwelt*)[7], but also to the world in which the producer and the user live.

What is apparent here is a reciprocal relation between the production and the use of a work. The manner in which the production of the work is ordered governs the use of the work; the rules obtained with respect to the use the work is put to in turn directs the making of the work. In this sense, the reciprocal efficacy of useful thing and work in their mutual referencing constitutes poiesis. (cf. Bernasconi, 1994, pp. 6–7) For this reason, Heidegger can 'translate' poiesis with 'handiness' (*Zuhandenheit*). §15 of *Being and Time* concerns Dasein's mode of engagement with entities in its environment. As is well known, Heidegger's thesis is that this engagement occurs primordially through relations of use. The used object is pre-thematic, in the sense that in its use it does not appear as the thematic core of Dasein's activities. The thing so understood is a thing in the terms which Heidegger finds already in the Greek term 'πράγματα', that is 'that, with which one has to do in concernful dealings (*praxis*).'(BT, p. 64; H, p. 68)[8] According to Heidegger, however, Greek philosophy failed to see this manner of being of the thing. Rather, it saw πράγματα as a *mere* thing: Greek philosophy ignored the equipment character of the thing. This is significant because the central claim of Heidegger's destructuring of the history of Western ontology is that the basic experience of Greek philosophy is that of making. The motivation for Heidegger's destructuring is that Western philosophy has for the most part hidden its original experience.

Heidegger defends the thesis, that the fundamental experience of Greek ontology is that of poiesis, through a discussion of Plato.[9] For Plato, ειδός (*Aussehen* – 'appearance') is based on μορφέ (Heidegger translates with *Gepräge* – 'form'). The essence of the thing is its appearance, the *how* of its appearing. This foundation relation is the opposite of what a philosophy that began with pure perception would understand. In perception, we encounter entities first in their appearance and then we come to their form. The fact that the Greeks – especially Plato – understood this in reverse, shows that they began from another experience. For the Greeks, that which *is* is moulded according to a model, which it must resemble. This, Heidegger points out, is the structure of production. The product 'owes' its being to its model, which is prior to the product. In the model, we find the 'what' of the product, its essence. The product leads back to that which was. In fact – and this thought has been familiar to us at least since Plato – it leads back to that which *always already* was. That which was always and always will be characterizes for the Greeks being as such. This does not mean that being was understood without relation to time; on the contrary, this definition of being is understood on the basis of time, time namely as presence.[10] Such an understanding refers to a past, which is contained in the present, as occurs in the case of the work produced.[11]

Through this destructuring, Heidegger wishes to return behind the forgetting of poiesis, in order to bring the equipment character of the thing to light. Implicit here is the question as to how theory arose together with the rise of the ontology of the thing.

In order to answer this question, we need to examine more fully Heidegger's understanding of the useful thing. '*Zeug*' is used in German either with other descriptive terms – '*Screibzeug*', literally writing equipment (typewriter); '*Fahrzeug*', motorized equipment (vehicle), etc. – or on its own. On its own, it is used in an undifferentiated way to refer to things, and could be best translated as 'stuff'. Stuff is precisely that which withdraws or remains invisible to theoretical contemplation; it is that without which something is and yet remains ineffable in that thing. It comes into appearance precisely as the stuff for writing, driving, etc. This coming into appearance of stuff as something to be used is the structure of the 'in-order-to'. The latter structure is a totalizing one: one useful thing can only appear within a totality. Every useful thing refers beyond itself to an encompassing order – a 'context of reference'. A useful thing refers immediately to a multiplicity of pieces of equipment, but such reference is subsequent to the order itself. Prior to the single useful thing, an ordering totality (the equipmental totality) is discovered. (cf. BT, p. 64; H, p. 68) The order is *prior to* the ordered.

The phenomenologically relevant question then is how this priority comes to appearance. It can only do so through an engagement with equipment which is true to its equipmentality. That is to say, the phenomenological analysis must begin with the use of the useful thing. Heidegger's famous discussion of the hammer and that act of hammering is an attempt to bring phenomenological rigour to the analysis of action. The hammer refers to a purpose or goal (*Zweck*) outside itself. As such, the equipment is inconspicuous; in using it, we look away from it and 'it is primarily not there as such a thing, but rather as a tool [*Werkzeug*], as *stuff for* [*Zeug zu*], which *is used*.' (Heidegger, 1985, p. 191; GA 20, p. 259 [translation modified]) So, in its being as equipment, the tool is disclosed its being as stuff for use, but as such is precisely not visible, but rather apparent as inconspicuous. At work, the worker is engrossed in his work: he forgets his environment. This forgetting has its own chronology, a chronology which, however, can be beset with crisis. Much later in *Being and Time*, Heidegger tells us that, in pursuing its concerns, Dasein is awaiting (*gewärtigend*), retaining (*behaltend*) and making present (*gegenwärtigend*). (cf. BT, pp. 373–7; H, pp. 406–11) These relations of Dasein are concretely expressed when its says 'then' (*dann*), 'before' (*damals*) and 'now' (*jetzt*). What is significant here is that, according to Heidegger, the 'making present' has a priority in concern (*Besorgen*). Implicit in 'before' is a 'not any more now" and in the 'then' a 'not yet now'. The now is always expressed in these temporal terms. The now is not a point but a transition (*Übergang*) (Heidegger, 1988, p. 248; GA 24, p. 352), something which Aristotle saw clearly. Against Bergson, Heidegger argues that this transition or movement cannot be understood in spatial terms. Movement, for Aristotle, is fundamentally a becoming other, which Heidegger names 'stretching' (*Dehnung*). (Heidegger, 1988, p. 242; GA 24, p. 344) The now is not a boundary because it means a 'stretching' from something to something.

In this stretching, there is a continuity, a constancy, which makes movement possible. (Heidegger, 1988, p. 243: GA 24, p. 344) The 'having-held-together-in-itself' (*In-sich-zusammengehaltene*), i.e. the constancy, has a unity. But there is in the now also a dispersal, a being rent asunder. The now becomes not any more now and is not yet now. The now is always an other. (Heidegger, 1988, pp. 247–8; GA 24, p. 350) The now is, in a certain sense, always the same, and, in a certain sense, never the same.

The now has, thus, a dimensionality which is irreducible to quantity. The now in which a shoemaker works on his shoes in the workshop is an other now than that which he spends in a bar drinking a glass of beer. This difference is not simply one of different positions of the now in the temporal sequence. Rather, every now is a 'now for ...' (*Jetzt, da*): hence, to every now belongs significance. The totality of relations between useful things Heidegger names 'significance (*Bedeutsamkeit*)' (cf. BT, p. 81; H, p. 87). Significance and the world of Dasein is constituted through the structure of handiness. In the same way as a useful thing in isolation from the totality in which it belongs is strictly speaking nothing, so too for a now in isolation. The now *is* only within an encompassing order of time which has the character of significance. This for Heidegger is 'world time'. It is the time of the 'work-world'.

As handy, the tool is structured as an 'in order to'. There is a time to hammer. The shoemaker works with a hammer, which is so crafted in order to produce a shoe. The time of this work is the time of handiness. 'In each case, things at hand are suited and unsuited'. (BT, p. 78; H, p. 83) The hammer is suited or unsuited for the task; similarly there are suitable and unsuitable times. (cf. Heidegger, 1988, p. 262; GA 24, pp. 369–70) Each time is a time in order to ..., a time to do this or that. There is a right time and a wrong time. These times refer to an encompassing order in which such a difference has significance and meaning.

We need to note here that the distinction which has been made up to this point between chronological and kairological time –namely, the former is to be understood quantitatively, and the latter qualitatively – needs to be relativized. Chronological time is only in abstract or formal cases purely quantitative. One reckons with time, but time does not in consequence fall into the indifference of an objective measure. The meaning of chronology is precisely that it has a sense and a directionality, and as such some times – chronologically understood – are more significant than others. In chronological time also there is a right time, something which up to this point we have encountered for the most part kairologically.[12] At this point, we need to ask how the possibility of kairos emerges in the chronological.

Everything which has been said of the chronology of hammering is only there implicitly. As long as the shoemaker is at work on the shoes he knows nothing of the chronology of his work. The hammer is inconspicuous, and with it time and the world. As with the useful thing, so too time slips into the background. At work, the shoemaker takes the time which he needs, and then forgets it. The order of the workshop is not disturbed; it remains taken for granted. It has a certain rhythm, a certain chronology of use. But this order of concern *can* suddenly be disturbed. Heidegger speaks of three possibilities of disturbance: the useful thing is unusable or it is lacking, is missing or something which is unhandy lies in the way. (cf. BT, p. 69; H, p. 73) In any of these cases the work cannot proceed. The order of the workshop is at once no longer self-evident: the hammer is damaged and for the first time becomes conspicuous and 'when we notice its unhandiness, what is at hand enters the mode of obtrusiveness (*Aufdringlichkeit*).' (BT, p. 69; H, p. 73) The leather is too tough and for the moment is not usable. The continuity of the work is interrupted. The handy thing can through this interruption appear in its mere presence as an objectively present thing (*vorhandenes*). Heidegger speaks of the shift from the mode of handiness to that of objective presence as a transformation (*Umschlag*). (cf. BT, p. 330; H, p. 361) The occasion for this transformation lies in a rupture (*Bruch*) in the context of reference. (cf. Heidegger, 1985, p. 189; GA 20, p. 256) Through this rupture, as we have seen, 'circumspection crashes into emptiness'. (BT, p. 70; H, p. 75 [translation modified]) But this rupture allows handiness to appear in its worldliness. The theoretical abstraction which thematizes the thing as objectively present is based in a forgetting of this rupture (but not in a disappearance of concern). (cf. BT, § 69b) The circumspection characteristic of handiness does not disappear with the disturbance caused by the breakdown of the useful thing. It is rather the case that, due to this breakdown, the useful thing is unusable for the purpose to which it had been applied. The reference of the 'in order to', which is normally inconspicuous, becomes obvious precisely because the referential relation can no longer be fulfilled. This lack of fulfilment brings the fulfilled, and in that sense perfect, context of reference into view.[13] That which comes to light here is not anything new, but rather 'the equipmental context appears not as a totality never seen before, but as a totality that has continually been seen beforehand in our circumspection. But with this totality world makes itself known (*meldet sich die Welt*).' (BT, p. 70; H, p. 75 [translation modified]).

Here, we encounter a kairological moment. In relation to a handy entity, in relation thus to poiesis, we can reach a moment in which a rupture occurs, a moment which cannot be calculated in advance. While orientating ourselves to the time of the workshop, suddenly there is a break in normality; circumspection crashes into emptiness. Emptiness is dark, is impenetrable for sight. In this darkness, circumspection itself becomes for the first time conspicuous. For the first time, it becomes possible to speak of circumspection in abstraction from dealings. A self-reflexivity in circumspection emerges, on the basis of which theory is possible. At the same time, 'world' announces itself in this darkness. This does not mean that world can be approached in circumspection. Only entities can be accessed in circumspection and world announces itself in the lighting up of darkness. It is this lighting up which makes circumspection first possible and in this way world is always already disclosed for circumspection. But world cannot be a theme for circumspection because the latter only occurs in world, only occurs on the assumption of world. Circumspection, hence all engagement with handy things, already assumes an order of reference. Only as long as world remains unthematic is the handy worldly, that is ordered in the context of reference. With the lighting up of world, comes a deworldling (*Entweltlichung*) of the handy thing: 'When the world does not make itself known, that is the condition for the possibility of what is at hand not emerging from its inconspicuousness.' (BT, p. 70; H, p. 75) This all sounds paradoxical: world is the lighting up, yet it appears precisely in darkness. World announces itself precisely as not present. It is rather to be traced in absence. (cf. Heidegger, 1985, p. 189; GA 20, p. 256).

This absence is not founded in the future; what becomes obvious is that the *world was always already there*. World opens itself 'behind' Dasein. In the having-been of this world, the poietical aspect of time discloses itself. Heidegger recognizes this as the 'aprioristic perfect' (cf. BT, p. 79; H, pp. 85, 441). It is here that the eternal announces itself in the poietical temporality. World is the order of handiness, which is always earlier than any handy thing. It forms an order which is that which pre-forms the handy thing. The 'in order to …' (*wozu*) and the 'for purpose of …' (*dazu*) are already pre-formed in this order. The useful thing has its significance and meaning only through the totality of reference; the rupture of this totality allows the world to appear as that which comes before, indeed which always comes before it – the world as eternity.

There is an obvious rejoinder to this interpretation: the world is grounded in Dasein and to the extent to which Dasein is not eternal, so too the world cannot be eternal. But in response it must be first pointed out that Dasein is conditioned by its understanding of being. (cf. BT, p. 10; H, p. 12) Understanding has the existential structure of projection. Projection is constitutive for being-in-the-world 'with regard to the disclosedness of its there (*Da*) as the there of a potentiality of being' (BT, p. 136; H, p. 145) Heidegger suggests here that projection needs to be understood on the basis of truth understood as unconcealment. I will explore the question of truth in some detail in Chapter 3. For the moment, what is important to see is that, in respect to handiness, we are speaking of one way of truth, that of poiesis. Dasein is conditioned through its poietical relations, which does not mean that it is conditioned as an entity: the poietical Dasein itself as an entity is not eternal.

It remains to be asked whether poiesis can be understood other than in the Greek way, which has been assumed up to this point. Might we not also understand poiesis as the bringing forth of the *new*? If, for example, we think poiesis, according to the model of the will – indeed the model of the divine will of a creator God as in Christianity – do we not end up with another concept of poiesis? For the moment, I have to leave this question hanging; I will return to it in Chapter 4. For the moment, I wish to press on with the interpretation of *Being and Time* and show how the poietical account conditions his analysis of temporality in that work.[14]

II Mood and mortality

If it is the case that Dasein discloses itself in understanding (cf. BT, p. 134; H, p. 143), and if it understands itself first in the modus of handiness (as maker), would we not be justified in saying that it is in its being a handy thing? Heidegger's distinction between authenticity and inauthenticity has the systematic purpose of foreclosing this possibility. Dasein is authentic when it fetches itself back from fallenness and understands its own being.[15] This self-understanding does not, for Heidegger, arise out of introspection, but rather out of the respective relations of Dasein. The truth is not 'in' Dasein, but rather Dasein is 'in the truth'. If it is the case that Dasein finds itself in mood in its transition to authenticity, and that it should be authentic as being-towards-death, we need to inquire as to the truth of mood. To understand this truth and the temporality of mood, two operative logics – those of failure and of reciprocal relation – need to be explored. These are already implicit in the principle of order of *Being and Time*, which I discussed in Chapter 1. That which fails allows the order which lies at its basis to come to appearance. What comes to appearance here at once depends on the essence of the failing order and conditions the transformation in and of that order. Between the order that fails (and hence passes) and that which emerges there exists a reciprocal relation.[16] These two logics in their interrelation correspond to the kairos; in the kairos, one order collapses, and at the same time a new one emerges.

II.1 Angst and handiness

For the most part, Dasein is engrossed in its concerns – in its work or in its engagement with things of use. For the most part, it busies itself with that which is at hand and handy. But Dasein is for Heidegger in its authentic being neither at hand nor objectively present. (cf. BT, p. 134; H, p. 143) Its being is, however, disclosed to Dasein only in moments of vision (*Augenblicke*), which occur for Dasein in its affective state (*Befindlichkeit*). The affective state of Dasein is manifest concretely in specific moods. Such moods for Heidegger are either affects of its engrossment with entities, as for example is the case with fear, or disclose for Dasein its ontological state, as is the case with Angst.

The analysis of Angst in *Being and Time* has the same structure as that of the breakdown of the handy thing. The analysis of Angst begins in the same place with Dasein's engrossment with the handy thing. This engrossment is disrupted by a failure, which can occur at a point in time and cannot be foreseen. 'Angst can arise in the most harmless situations.' (BT, p. 177; H, p. 189) While the analysis of Angst concerns the worldhood of Dasein, not the worldliness of the handy thing, the inner relation between both is crucial with respect to the difference between praxis and poiesis. The worldliness of the handy thing is based on Dasein, which engages with the world in the mode of handiness. The question then is to what extent the poietical experience of such a mode of handiness forms the basis of the analysis of Angst.

In terms of its chronology, the same account can be given of Angst as of the experience of the disruption of handiness. In the case of Angst admittedly, not only the handy thing, but the 'the totality of relevance discovered within the world of things at hand (*Zuhandene*) and objectively present is completely without importance.' (BT, p. 174; H, p. 186) For this situation to arise not one handy thing needs to fail. It is rather the case 'that everything at hand and objectively present absolutely has nothing more to "say" to us.' (BT, p. 315; H, p. 343) When a handy thing fails, it nonetheless still has a relevance (*Bewandtnis*) for Dasein. But in the affective state of Angst such relevance falls away; Dasein finds itself alone, as a *solus ipse*. (cf. BT, p. 176; H, p. 188) In this moment of failure, Dasein is brought before nothingness (*das Nichts*). It is not only the case that Dasein in its circumspection crashes into emptiness; nor is it simply the case that Dasein no longer knows how it should proceed with the handy thing; it is rather the case that Dasein finds itself in a situation in which it has no orientation. This lack of orientation Heidegger terms 'unhomeliness' (*Unheimlichkeit*). The order of reference with which Dasein is familiar, which indeed makes up its everyday life, breaks down. (cf. BT, p. 176; H, p. 189) In the midst of this breakdown, Dasein is still in the world, but is not at home in the world. Dasein finds itself anxious before nothingness, that is, before no inner-worldly handy thing.

Through unhomeliness, Dasein is immediately brought before itself. In Angst, Dasein is individuated, because it finds itself alone. For Dasein, the handy things no longer have any relevance. It is for Dasein as if it is in an alien world: what previously was familiar now has no more meaning or significance for it. The old order has become alien to it. But this alienation is no accidental occurrence: unhomeliness is more original than familiarity. Dasein is not at home in the world of handy and objectively present entities. In Angst, it becomes clear to Dasein that its being is other than what it is familiar with – the handy things in the order of handiness. While the failure of the tool discloses the worldliness (*Weltmässigkeit*) of the handy things (cf. BT, § 16), Angst reveals Dasein in its worldhood (*Weltlichkeit*) and as having an order prior to that of handiness. (cf. BT, § 40)

Dasein finds itself in its temporality between chronos and kairos. Angst brings this 'between-space' to appearance. For Heidegger, the present of Angst 'holds the moment of vision in readiness for a leap [*auf den Sprung*], as which it, and only it, is possible.' (BT, p. 316; H, p. 344 [translation modified]) Dasein is brought to the point of rupture, but it requires resoluteness (*Entschlossenheit*) to make the leap over this breach. This amounts to a radicalization, but not a new step beyond the analysis of equipment. In both analyses, Dasein comes to a rupture that opens up not a future, but a having-been world, a world which has always already been. In both cases, it is the having-been (*Gewesenheit*) which announces itself. In Angst, Dasein finds itself as thrown, finds itself as delivered over to a having-been, for which it can bear no absolute responsibility. Dasein encounters here a past which can never have been its past, a past however which it must always assume.

The thing as handy and as objectively present says nothing to Dasein when Angst arises in it. This, for Heidegger, also extends to other Daseins: 'The "world" can offer nothing more, nor can the Mitdasein of others.' (BT, p. 187; H, p. 187) Simultaneously, the relations of being with things and with other Daseins fail. Those relations can say nothing to Dasein, nor offer anything to it. This failure of other Daseins is mentioned by Heidegger almost as an afterthought. This indicates the poietical experience underlying the analysis. Angst is situated by Heidegger in the world of handiness, the work-world. 'Along with the usability of the work, the work-world at the same time *appresents* the world in which users and consumers live, and in this way it appresents them too.' (Heidegger, 1985, p. 192; GA 20, p. 261 [my emphasis]) Being-with of the work-world is the being-with of users and consumers. In the case of Angst, the work loses its meaning and significance, so too do the users and consumers. In a world in which the other is 'appresented' in this way and is not there – for example, as friend or as conversation partner – Dasein is alone; as a maker, it does not need the other. (cf. Arendt, 1998, pp. 153–9) Dasein is individuated in this way because, and so long as, it lives in the world of poiesis. It is this world, which always already was, which makes up the having-been of such a Dasein. Indeed, Dasein is 'thrown' into such a world.

Dasein is understood by Heidegger as a 'thrown-projecting-being-in-the-world'. Dasein's being is its understanding of its own situatedness as worldly. The analysis of Angst is concerned with the manner in which Dasein encounters its own being-in-the-world. The world in which Dasein finds itself is a work-world. Dasein can only experience itself as a totality in such a world. This is so because, above all else, Dasein must be able to experience itself as individuated and alone in order to experience its own totality. To the extent that the world of users and consumers is only appresented, the work-world is a world of loneliness for Dasein in its relation with itself, allowing it to understand itself as a totality. Furthermore, this world is that which always already was, and in such a world the having-been has priority.

This account appears to contradict Heidegger's analysis of 'care' (*Sorge*) as the structure of the totality of Dasein: 'The being of Dasein 'reveals itself as *care*' (BT, p. 171; H, p. 182 [emphasis in original]). Furthermore, Franco Volpi has argued that care can be reconstructed as praxis not poiesis. (cf. Volpi, 1988) There is, however, an ambiguity in the structure of care, which allows it to be understood under two aspects, those of poiesis and praxis. To show the poietical aspect of care, it is necessary to turn to its temporal sense as being-towards-death.

II.2 Death and the future past

In the being of handiness, the past has a priority, which leads to the primacy of eternity over time. The temporality of Dasein is, however, finite, a finitude which Dasein experiences in Angst. In Angst, there occurs a breakdown of order of poiesis, which leads to a priority of the past. Through this breakdown, and in the context of this priority of the past, Dasein encounters the question of its own totality. As we have seen in Chapter 1, for Heidegger, this totality is based in being-towards-death. This totality is only to be experienced in the world of poiesis, and as such characterized through a *chronological* temporality.

In being-towards-death as 'anticipation', as literally 'running forward' (*Vorlaufen*), we can glimpse the futuricity of the temporality of Dasein, which is phenomenally indicated first in understanding. In Angst, Dasein finds itself as thrown towards death, but at the same time understands itself as projection, as running-forward, as always ahead of itself. But in understanding this being always ahead of itself in terms of death, Heidegger in fact accounts for Dasein not as future, but as *having-been*, and, in this way, handiness gains a primacy.

Heidegger thinks of death as an end. As long as Dasein lives, it is not yet dead, the end is not yet, it is 'lacking'. There is, in other words, something 'outstanding' (*Ausstand*) for every Dasein. What is outstanding is based on belongingness as that which is not yet available. Hence, Heidegger says 'to be outstanding means that what belongs together is not yet together'. (cf. BT, p. 225; H, p. 242) Ontologically, what is in question, here, is the unhandiness of those parts of life which have not yet been brought together. The entity for which something stands outside has the manner of being of the handy thing. Dasein, however, is not such an entity; it is rather a being *towards* the end. Dasein is to be understood as a movement towards the end. Death is not given, but is rather a possibility, the possibility of the impossibility of Dasein. Dasein is a totality in this movement because it exists in the running forward to its utmost possibility.

Death is the utmost not-yet. It has the character of something 'to which Dasein relates'. (BT, p. 231; H, p. 250) Heidegger's use of the expression 'what-for' (*wozu*) needs to be noticed here. According to Heidegger, the totality of relevance is

> "earlier" than any single useful thing … The total relevance itself, however, ultimately leads back to a what-for [Wozu] which *no longer* has relevance, which itself is not an entity of the kind of being of things at hand within a world, but is a being whose being is defined as being-in-the-world, to whose constitution of being worldhood itself belongs. This primary what-for is not just another for-that as a possible factor in relevance. The primary "what-for [*Wozu*]" is a for-the-sake-of-which (*Worum-willen*). (BT, p. 78; H, p. 84)

In this case, it is clear that the 'what-for' is death, as the possibility of the impossibility of Dasein. Dasein encounters its worldhood as 'what-for' out of the experience of handiness. The connection between Angst and death arises precisely in this context. Angst as such gives Dasein no hint of death or mortality – if, for example, it was possible to reverse the ageing process[17], as some recent research is aiming to do, such that it would be possible for people not to be essentially mortal, this would do nothing to eliminate Angst nor the sense of nothingness.[18] In Angst, Dasein encounters nothingness. Even if this nothingness is the nullity of its existence, this cannot without further ado be interpreted as death or mortality. It is precisely through a particular, but not compelling, interpretation that such nullity and death are brought together. Even when Dasein understands Angst in a state of resoluteness, there is no compelling reason to say, with Heidegger, that Dasein understands itself as being-towards-death.[19] To understand the connectedness of death and Angst in Heidegger's analysis, it is necessary to inquire as to the character of the world of Dasein. It is the *work*-world. In it, Dasein understands itself out of the experience of handiness. Dasein encounters Angst and mortality in the midst of its work. In Angst, the world of the handy things comes to lack all meaning and significance; in death, Dasein encounters the final what-for as for-the-sake-of-which that it cannot find among handy things. But what it finds is, so to speak, so conditioned by the '*via negativa*' of handiness that Dasein understands itself as *not* at hand. We know from Heidegger himself the danger that such negative paths lead to an understanding coloured by that which has been negated. This danger is precisely that of which we spoke under the title of the 'logic of reciprocal relations'.

For Heidegger, the anticipatory running before itself into death (*Vorlaufen in den Tod*) makes the ahead-of-itself of the being a whole possible. But, as Heidegger himself has shown, such being-ahead-of-itself (*Sich-vor-weg-sein*) in the existential of understanding arises from a future understood as an open possibility. In projecting a possibility, Dasein is concerned with itself; in such projection it comes back to itself. But such a coming back to itself can never be final. If it were final, then Dasein would have actualized itself, and as such would relate to itself no longer as projecting but as the already having been ultimately projected.[20] The futural direction of understanding has the structure of a circle, but that circle can never be finally closed, and rather remains the possibility of its own continual re-opening.[21] If, however, the highest possibility of Dasein is founded ontologically in a running forward in anticipation towards death, then the possibility of Dasein comes under the sway of actuality.[22] Death becomes not a possibility to be projected, but a projected actuality which is not yet. Heidegger can only totalize the movement of Dasein, because he has surreptitiously gone beyond the threshold of the (actual) death of Dasein[23] and observed the movement of Dasein retrospectively from its outer boundary. David Carr has quite perceptively interpreted the futuricity of being-towards-death as a 'future-*past*'. To consider a life as a totality means to consider it *as if* it was already past.[24] If we are to interpret Dasein as being-towards-death, we can so only from the position of 'existentiell neutrality' (BT, p. 231; H, p. 248). This does not just mean that it has no existential point of guidance, but also that it must be performed outside the movement of the authentic temporality of Dasein. It must consider the temporality of Dasein from its end, which means that the analysis itself contradicts the authentic movement of temporality which it is attempting to bring to light. As ahead of itself, Dasein has an authentic futural sense, but if this has its final ontological confirmation in being-towards-death, then Dasein is no longer understood as futural, but as already having-been. What thus comes to appearance is not the arising of a meaning to come, but an always already having-been significant. In other words, the account of being-towards-death does not in fact interpret Dasein from its mortality, but rather from an anticipated ideal look back over its own existence, at which point a completed chronology of the life is possible. From the perspective of this backwards glance, there can be no genuine kairological rupture. Dasein as being-towards-death is the *auto-biographical* – the auto-chronologizing – Dasein par excellence.

How can such a backwards glance be possible at all? With this question, we encounter the problem of the other.

Hans Blumenberg maintained against Heidegger that the self's own death can only be experienced through the other. I have no unmediated knowledge of my death, or for that matter of my birth. 'The proposition, all people are born and must die, is the result of inter-subjective experience', he says. (Blumenberg, 1986, p. 91) Death (and birth) occurs only in the public space of everyday being with one another. According to Blumenberg, Heidegger makes finitude into an immediate content of conscious awareness. Against this, we must hold fast to the realization that Heidegger is concerned not so much with consciousness as with action, and the putative immediacy of contents of consciousness is unavailable to action. In the case of death, Heidegger begins from the everyday experience of death and sees in that experience both the threat of death and the need for security from death. Heidegger would not contest that the first knowledge Dasein has of this is obtained in the inter-subjective space. He attempts, however, to move from this ontic level to the ontological (which for Heidegger is 'earlier' than the experience of particular others). But the thrust of Blumenberg's objection remains, namely, whether it is possible to engage in such an ontological account without reference to the inter-subjective presuppositions of the experience of death and finitude.

According to Heidegger, we do not experience the loss of being of another deceased person. (cf. BT, p. 222; H, p. 239) We do, however, notice a loss. Somebody is no longer. She no longer speaks to us. Her past actions may continue to affect us, but she is no longer a being-with us. Her life is past; we begin to speak of her in the past tense. We begin, indeed, to tell stories about her life. These stories have an element of completion about them. We seem to be in a position to decide whether her life was happy or not, successful or not. Only after death does this seem to be possible.[25] The deceased no longer 'moves' herself in the sense of her existentiell movement. In ontical and everyday terms, we have here an exemplary form of Dasein's totality (*Ganzsein*). The question is now whether it is so also on an ontological level.

Dasein's being-towards-death concerns its own death; what is at issue is *my* death. But if Dasein is only made known of its own death through the death of another, then this difference between my death and that of the other is not as decisive as Heidegger assumes. The question arises then, to what extent we can really say that Dasein's being with others fails in respect to death. Heidegger stresses that no one can take its death from an other[26]; but the above analysis has shown that Dasein experiences its possibility of its totality through the demise of another. Dasein is individuated in Angst, in the sense that the other fails it in its Angst; but this individuation can only then be interpreted as being-towards-death when it experiences the death of another. In analogy with the demised other, Dasein can understand itself *as if* dead and as such can consider itself in its totality.

What is at issue here is Dasein's understanding of its own being. Dasein is authentic when it understands itself not on the basis of its experience of handy or objectively present things, but rather from its own self. If such a self-relation does not genuinely arise from introspection (cf. Heidegger, 1988, p. 159; GA 24, pp. 225–6), how else does Dasein come to such self-knowledge if not through the 'mirror' of another. The metaphor of a mirror is itself suggested by Heidegger's own account of truth in terms of a metaphorics of light and of reflection as 'to break at something, to radiate back from there, to show itself in a shining back (*Widerschein*) from something.' (Heidegger, 1988, p. 159; GA 24, p. 226) Furthermore, he characterizes self-knowledge in terms of transparency. (cf. BT, p. 137; H, p. 146) In the face of which other, though, can Dasein find a mirror for itself? The relation to the other which such an analysis of dying assumes is a relation to someone who does *not* speak to me. Such an other may have needs, but is not a partner in conversation. We encounter this other which does not speak to me in the deceased person. This experience of not being addressed by the other, an experience of the other as beyond the reach of conversation, as other than what is present to me in my preoccupations, is one which is reflected for Dasein in its work-world. In its work-world, the other is not present as living, but as an other, which, though it can address me, does so only to interrupt the flow of my work. Furthermore, it is an other whose needs and dimensions are for me already complete, already there written into the completed product of her use. Such an other is present as having already completed her happy and successful action with the product of the workshop. Such an other mirrors to Dasein its own totality and itself as a being in relation to its own end, understood as death.

Heidegger wishes to transcend Dasein's particular ontical situation, but the ontological analysis remains that of a specific Dasein, namely Dasein as maker. This makes sense of the fact that, in Heidegger's analysis, Dasein comes to be understood as having-been. As individuated, Dasein finds itself in its work-world. Being-towards-death is one possible way of understanding the being-towards-an-end – finite Dasein. It is the interpretation of Dasein which lives in a work-world. Thus, the world of praxis is left out of view and the 'futuricity' of Dasein is undermined.

Dasein understood as being-towards-death is thought neither praxically nor politically, but rather poietically. In this view, the having-been has priority. Furthermore, the finitude of time is not understood in a radically kairological manner. The order of time in this poietical understanding is always already there. What this means is that time is only the occasion to become aware of this order, which is being intended not as temporal, but rather as eternal. This order does not arise for Dasein, rather in the moment Dasein understands the order of the world in such a way as if it had always already been there.

Nevertheless, such temporality is not purely chronological because the being towards an end remains related to the moment. The 'when' of death remains undetermined and cannot be calculated. Death is possible in every moment. (cf. BT, p. 238; H, p. 258) This experience of the *threat* of death can serve as a starting point to rethink temporality as praxical not as poietical.

III The priority of futuricity in the temporality of Dasein

In this section, I will attempt to give a reading of 'care' (*Sorge*) running counter to the poietical interpretation sketched above, in order to draw out its praxical characteristics. The priority of an open future in Dasein's temporality is closely connected to such characteristics. The temporality in question here is kairological. This interpretation is not directed against Heidegger, but rather corresponds to the other aspect of his deeply ambiguous presentation of Dasein's temporality. This aspect is a kairological one, which comes to expression out of the experience of Dasein as being in praxis.

'"Future [*Zukunft*]" does not mean a now that has *not yet* become actual and that sometime *will be* for the first time, but the coming [*Kunft*] in which Dasein comes toward itself in its ownmost potentiality-of-being.' (BT, p. 299; H, p. 325) Heidegger is distinguishing here between the chronological future, as a now which is not yet, and the future in the sense of a kairological order. Understood chronologically, the future is, strictly speaking, already having-been. It has not yet appeared and is not yet past, but is nonetheless homogenous with the past time: the future will have been. But, as the above quotation makes clear, the future for Heidegger is a coming. The coming is to be thought as possibility; it is, strictly speaking, not actualizable. The coming lies always beyond Dasein's reach. It is the possibility of an impossibility, namely the impossibility of Dasein coming to itself. As to-come, as futural, Dasein cannot close the circle of its being because its being has the character of being possible. (cf. Heidegger, 2001, pp. 3–4; ZS, pp. 3–4) Heidegger thematizes this impossibility of closure as the ecstases of temporality.

The temporal modi of future, having-been and present are related to one another, or rather *are* only in this interrelation. This is a way of openness. There is in this interrelation a certain explosiveness, a bursting forth through which temporal distance is first possible and in which the world first emerges. This bursting forth frees entities in their being open towards Dasein. This is the origin of transcendence, which will be an important theme of the next chapter. *'Temporality is the original "outside itself" in and for itself.'* (BT, p. 302; H, p. 329 [emphasis in original; translation modified]). In this sentence, we can find the basic principle for Heidegger's concept of kairological time. Temporality is a movement, indeed movement as originating: movement as the making possible of relations to anything at all. As such, time is not an entity, has no determinations, is rather 'outside itself'. Anything determined or determinable stands in a context in which it is a part; but temporality is the last horizon, the horizon, namely, of being. Despite its indeterminateness, temporality has the character of the 'in itself', due to its unity. 'Temporality is not … an entity that first emerges from *itself*; its essence is temporalizing in the unity of the *ecstasies*.' (BT, p. 302; H, p. 329) This unity is of the three temporal modi, which are co-original. This means that the movement of time *is* only in its bursting forth in three directions.

Only thus is transition (*Übergang*) possible. This unity is that of the 'towards itself', 'back to' and 'letting something be encountered'.

These three prepositions – towards (*zu*), to (*auf*) and together with (*bei*) – indicate the three sided unity of the 'ahead of itself' (*sich vorweg*), the 'already being in' (*schon-sein-in*) and 'the being together with' (*Sein bei*) in care. But this unity is not immediately or generally accessible; it comes to appearance only momentarily. Heidegger writes: 'Original and authentic temporality temporalizes itself out of the authentic future, and indeed in such a way that, futurally having-been, it first *awakens* [*weckt*] the present.' (BT, p. 302; H, p. 329 [my emphasis; translation modified]).

'Awaken' is a remarkable term here. It is one which we find in many places in Heidegger's lecture courses from the 1920s.[27] That which is awoken is both there and not there. This phenomenon Heidegger calls 'being-away [*Weg-sein*]'. It is part of the way in which Dasein is. Dasein is simultaneously there and absent. This simultaneity is a condition in which Dasein can step out of its temporality in the chronological sense. Only because Dasein can experience this split, indeed is this split, can it experience its temporal ecstases as a unity. In this unity, Dasein is at once its having-been, its future and its present. It is these only because 'a human being – insofar as he or she exists – is, in his or her being there, also always and necessarily away in some manner.' (Heidegger, 1995, p. 63; GA 29/30, p. 95) and as such can be awoken.

The capacity to be away indicates the futuricity of Dasein. A philosophical approach that stresses the present and presence will consider being away, hence 'absence', as a lack in the attentiveness of consciousness. In such a view, the human being has only the possibility to know the truth when he is present to an object. If, however, the human is both there and not there, then it is not possible, not even in an ideal case, for the human to be *only* there, *only* present. But Dasein is simultaneously there and not there. Even alone, it can never be fully present to itself. This is what makes it possible to awaken Dasein. Waking is a coming-to-itself of Dasein out of its not-Da-sein. As such, awaking means the opening up of the future.

The moment of waking is not merely the point in time in which waking occurs. Rather, original and authentic temporality wakes the present. The present is transformed in the moment through the glimpse of possibility as 'coming', becomes as such not an object of concern but rather is understood as the time of waking.[28] Waking brings possibilities which had not been revealed, but already were, to appearance; but that which is woken is not simply actualized, is not simply made into an object of observation, but rather is set in movement. (cf. Heidegger, 1995, p. 65; GA 29/30, pp. 97–8) This movement is that of possibility as the moment, as kairos. As such, it is a movement into a future which is not already sketched out in the past. It is a movement of Dasein as ahead of itself. In this situation, Dasein is not related back through failure to a pre-existing order that always already was. If the future is not already sketched out, then a new order can emerge.[29] Dasein that discloses this new order and which can respond to this new order is the acting, praxical Dasein. According to Heidegger, the being of Dasein is authentic action, praxis. (cf. Heidegger, 1984, p. 183; GA 26, p. 236)

In his lecture course on Logic from the winter semester of 1925–6 (Heidegger, 2010; GA 21), Heidegger points out to his audience that the seeing of the matter of his lectures cannot be *produced* through the lecture, but can only be *awoken*. The lecture itself is a form of communication. The mode of being of a communicating Dasein in its relation to its hearers is not a being-together-with (*Sein bei*), is not a function of concern, but is rather a being-with, a caring with or a caring for (*Fürsorge*) (cf. Heidegger, 2010, p. 187; GA 21, pp. 222–3). The distinction here is between that which is produced and that which is awoken. Awaking is not a matter of poiesis; it is in the realm of communication and of speaking and sharing together – of speaking and hearing, that is of being-with – that something can be awoken. In the work-world, things are inconspicuous, and when the handy thing loses its meaning and significance in Angst, Dasein can communicate nothing: the world is sunk into insignificance, but significance founds words and speech. (cf. BT, p. 82; H, p. 87) In Angst, Dasein is 'mute'. What is at issue here, however, is to open up a sight for that which calls to be communicated. The sight which is to be awoken cannot, therefore, be that of circumspection and cannot be a poietical capacity. To see those matters that are at issue in such speaking and hearing, what is appropriate is not the poietical circumspection, but, rather, considerateness (*Rücksicht*) and forbearance (*Nachsicht*). (cf. BT, p. 115; H, p. 123) What are these matters? For Heidegger, they are those in which Dasein understands itself. The structure of care is that of Dasein as an entity which understands. It is this 'matter', that of Dasein itself, which is invisible to circumspection.

In this, Heidegger hints at another non-technical possibility of theory. It concerns not so much the sight of dealings with useful things, but rather the sight for that which deserves to be communicated and shared. Such a theory arises out of the failure of communication. But if, in this failure, nothing objectively present and no eternal order comes to appearance, then this is not a theory of dying, but rather one of natal being.[30] This opens up the possibility of philosophy as a theory of the new, which must all the same remain praxical. That which is to be communicated could in traditional terms be called '*telos*', as in Aristotle, where happiness is the goal and purpose of human life and action. We must be cautious, though, in employing such a term, because the *telos* is the fulfilment of what is always already there; what is at issue here rather is the *emerging* of an order. This is the consequence of thinking the future as possibility in a radical sense.

Dasein *is* in its understanding of being. Understanding as self-projection is, for Heidegger, the authentic sense of action.[31] If Dasein is only authentic to the extent to which it understands itself out of the movement of understanding, then it is authentic only as an *acting* Dasein. This is already suggested in Heidegger's characterization of Dasein as existing in concern about its own being. This self-relationality of Dasein corresponds to the structure of praxis in Aristotle, according to which praxis has its telos in itself.[32] In an Aristotelian sense, Dasein as care is praxical Dasein. Nevertheless, Heidegger's account, while understanding Dasein in terms of such a self-relationality of praxis, does so in terms of a radically non-teleological ontology.[33] Telos is fulfilment, completion; in temporal terms, it is to be understood chronologically, in terms of a determinable beginning and end. If, however, the future of praxis cannot be enclosed in the 'circle' of Dasein, due to Dasein's being as possibility, then it must remain indeterminable and uncertain. The 'for-the-sake-of-which' (*das Worumwillen*) cannot be understood as a future goal.[34] Furthermore, teleology assumes that the end is already contained in the beginning. The latter rules out any radical novelty, which – as will be discussed in the next chapter – is rooted in a contingency that makes any derivation of the future out of the having-been questionable. If it is the case that there is a nothingness as an end, as a moment, in which at once the decline and the emerging of orders occur, or better as a moment between decline and emerging, then we can glimpse in 'care' a new logic of failure. This order takes Dasein beyond the everyday to the contingency of the new. At the same time, the future becomes a 'genuine' future and possibility is opened up, in which temporality is seen in its temporalizing between Dasein and in each case its distinct other Daseins with which it is.

What Heidegger discovered in care – according to this praxical interpretation – is not the being whole of Dasein, but rather its groundlessness, its abyssmalness. At the heart of praxis and in it alone, the uncertainty of the future becomes manifest. (cf. Held, 1993, p. 404) This uncertainty is not essential to Dasein as maker, but rather to Dasein in praxis, where it is *in the same world* with others (cf. BT, pp. 115, 351; H, pp. 123, 384) and not merely being-with in a derivative sense. Dasein's dependence on time is indicated in this uncertainty.

This dependence becomes clear when we consider that there is a 'right time' for action, that there are kairological moments, in which Dasein is called upon to act. In poiesis, we have no experience of such times (only of the time of inaction, where the tool breaks down): if something is to be produced, then the material is ready to be used; there is no time at which it *must* be used. The schedule of production is not determined by the material, but rather by the appresented world of the users and customers. There is no 'right time' in the work-world, at most a 'suitable time.'[35]

At this point, it is important to point out an apparent contradiction in the above analysis. I have already pointed to the fact that there are right times within chronology, and have done this precisely in the context of poiesis. However, such right or inopportune times are not determined by the work-world itself, but rather by the appresented world of the users and customers. The maker as such knows of no such right or inopportune times, these kairological elements form the dimension of intersubjectivity into chronology. Nevertheless, such times are not understood kairologically in the everyday world, but rather are interpreted in terms of a sequence of nows. (cf. BT, pp. 373–7; H, pp. 406–11; Heidegger, 1988, pp. 361–4; GA 24, pp. 369–74)

According to Heidegger's interpretation of Aristotle, for the latter, the time of action is itself an *ἀγαθόν*. The 'good' deliberation concerning praxis is accompanied by a proper reflection on time. The issue, here, is not a matter of the length of time spent deliberating: rather what is at issue here of the feeling for the right time to act. Thought kairologically, the right time is that of the moment in which an *order* collapses. The question then is what form of worldly order collapses in such moments.

The facticity of Dasein means that it is 'thrown' in a common world with others. This is the public world of the 'the one-self' (*das Man*). The latter is for the most part characterized negatively in *Being and Time*. However, if futuricity is to be understood in praxical terms, and if the future has a priority in original and primordial temporality, then the praxical world must have a primacy with regard to the work-world. If that is the case, then the public world is not simply that which is appresented in the work-world; it is its own space as the world of being-with in praxis. In other words, if the interpretation attempted here is correct, then an account of the political world must be implicit in *Being and Time*.[36]

IV Friendship and boredom

It seems without question that Heidegger understands the other for the most part in *Being and Time* in the context of the work-world. As he says, we 'meet them at work'. (BT, p. 113; H, p. 120)[37] As a result, it is the loneliness of Dasein rather than, say, Dasein as friend which is mostly in view. And yet, this very loneliness is constantly disrupted in the course of the analyses of *Being and Time*. This is so because Dasein precisely in its authenticity is understood as responsive, as listening and responding to that which undercuts its tranquilized state. Christopher Fynsk is guided in his analyses of *Being and Time* by this structure and suggests that we 'might recognize the other as providing the intervention necessary for drawing Dasein out of its subjection to the they'. (Fynsk, 1993, p. 48)

If Fynsk is right, then we need to read *Being and Time* as a whole in terms of its structure of responsiveness, and hence as a circular movement around an unfathomable origin. In that case, the friend is not one with whom Dasein can feel at home, but precisely the one who discloses to Dasein the abyss of its own being. The friend is mentioned directly only once in *Being and Time*: 'Hearing ... constitutes the primary and authentic openness of Dasein for its ownmost possibility of being, as in hearing the voice of a friend whom every Dasein carries with it'. (BT, p. 153; H, p. 163) It is not by accident that, on this one occasion, it is the voice of the friend which is mentioned – hence the friend occurs in the context of hearing and speaking. Friendship is not itself made thematic, and arguably this single reference should be understood in the context of the call of conscience, rather than in terms of a relation of friendship with another. It is, nonetheless, remarkable that Dasein which understands itself only in relation to other entities – self-understanding does not, as we have seen, emerge from introspection – should not be discussed in terms of its reciprocal relations with others. Nor can the issue of friendship be ignored simply on the grounds that it is irrelevant to Heidegger's concerns. If – as Heidegger states – Dasein is the ultimate 'for the sake of which', the question as to the 'in order to' of the useful thing for Dasein is unavoidable. Only the good of Dasein can count as such as an 'in order to'. But how is that good to be decided if not in community, that is in conversation with one another, in short, in some sense through a political decision? Such a decision is one which emerges from a certain understanding of the good for Dasein. Such an understanding is a once a self-understanding on the part of the individual Dasein, and a commonly arrived at understanding on the part of a particular community. It is in such a context that I wish to place the issue of friendship.[38]

According to Aristotle, friendship is made possible through self-love, but emerges at the same time through and due to limitations of the self. (cf. Aristotle, NE, 1169a25–35; cf. Gadamer, 1999; 1991) Due to these limitations, the self needs an other. If the human were perfect, he would – like a god – need no friends. Dasein is – as Aristotle described the human – always underway to itself, on the way to self-knowledge. For Aristotle, the friend acts as a mirror on this way, a mirror in which we recognize ourselves. (cf. Gadamer, 1999, pp. 138–9) In such a mirror, Dasein does not see the particularity of its own being, but rather the common, which binds it with the other. Friendship as such is precisely 'not a matter of one or the other' (Gadamer, 1999, p. 139), but rather lies between. The realm between is the common; it is a realm which we call 'community'.

Friendship requires time. (cf. Aristotle, NE, 1156b25). This is not the time of one or the other, but rather the time of both. It is not clock time, not time as measure. Friendship requires time not in the sense of an indifferent sequence of nows, but also not in the sense of an external coupling of mine and your time. Rather, it requires a common time that passes in harmony for both. In English, we have a felicitous term to express this, namely 'timing'.

Timing is a phenomenon of simultaneity. If, for example, two people arrive at an appointed place at the same time, we say 'that was good timing'. Timing can, however, have an erotic meaning: the bodily sensibility for the other, the feeling for time in the giving over of oneself to a rhythm, which is neither mine nor the other's, but rather has a power over both. What emerges here is not an identity of 'my' time and 'your' time. It is rather the case that 'we' find ourselves so bound with one another in this time, that it is the very rhythm of this time that makes our being together possible. Time temporalizes itself not 'in' me or 'in' you, but rather between us in a between-space[39], as a power between and over us.

While timing so understood has a clear erotic basis, such an erotic forms the basis not simply of friendship – *θιλία* – but also of dialogue more generally. The private space of the erotic gives a sense for the rhythm of timing, which finds public expression in friendship. That which is operative here is a power of time, felt most intensely in erotic encounters, but functioning in a mellower form in friendship and dialogue more generally. Every dialogue assumes this space of a 'between-time'. As I wander down the road lost in my thoughts and suddenly someone unexpectedly addresses me, this 'between time' becomes immediately conspicuous as that which breaks into my Dasein – occasionally as a threat. When someone addresses me in a foreign language, which I have not yet mastered, this experience is all the stronger: I experience the rhythm of this conversation itself as foreign and strange. Only slowly and perhaps stammeringly can I respond to this address. (cf. Gadamer, 1976, p. 16; 1993, p. 230) In friendship, we experience this between time, this power of time, but – when things go well – no longer as threatening or as strange. Rather, we find ourselves in tune with it and, through this, in tune with each other. The reverse side of this is, however, time as threatening and alienating.

Such timing can be made homogenous, routine and indeed mechanical. The continuity of life can be reduced to a pure chronology in which timing is manifest simply as mechanical routine. On the other hand, timing can be experienced kairologically as a moment of pure, and at its most intense, erotic harmony.

It is precisely this incompleteness of Dasein which opens up the possibility of friendship. Dasein as being-towards-death needs no friends, but in its encounter with the other Dasein experiences its boundaries. Friendship is a possibility – perhaps the possibility – of transcending such boundaries. In being in harmony with its friend, Dasein can see itself in the other, in which it finds itself mirrored back to itself. Such a harmony is nonetheless unstable; even the friend remains a stranger for Dasein. The other is always somewhere else and, even when timing occurs, in an other time. In this sense, the other fails in the sense of holding itself back, withdrawing from Dasein. This withdrawal of the other, even as friend, can be experienced in various moods. I wish here to concentrate on one such mood, namely, boredom. Boredom responds to withdrawal; to be bored is to experience the falling away of that harmony which characterizes the between-time. Furthermore, in the wake of *Being and Time*, Heidegger gives systematic consideration to this mood in his lecture course of winter semester 1929–30, *The Fundamental Concepts of Metaphysics*. (Heidegger, 1995, pp. 59–167; GA 29/30, pp. 117–249) It is in his analysis of boredom that Heidegger first explicitly discusses the 'power of time'. It is, as we have seen, in friendship that this power is most clearly manifest. It can manifest itself as joy in the harmony of friends, but through a process of alienation can emerge as threatening and oppressive.

Mood, Heidegger says in this lecture course, is the how of being with one another; it is, so to speak, the *atmosphere* in which Dasein finds itself. (cf. Heidegger, 1995, pp. 66–7; GA 29/30, p. 100) Mood is not a subjective reaction, but rather emerges between Dasein and an other. Heidegger also calls mood a 'medium'. (cf. Heidegger, 1995, p. 67; GA 29/30, p. 101) It is not made possible through the actions and sufferings of Dasein, but rather forms the medium or atmosphere in which these occur. Without moods, no interaction with others could occur because they disclose the world, which is always a world with others.

Boredom is a mood which brings Dasein, through the chronological passing of time in its 'naked succession' (cf. Theunissen, 1991, p. 304), to the moment of the kairos. Two main phenomena need to be discussed in this regard: 1) situation, and 2) the power of time.

1) 'Situation', which Theodore Kisiel characterizes as an 'old kairological concept' (Kisiel, 1993, p. 423), is used in distinction from location (*Lage*) as a temporal concept in *Being and Time* and in lecture courses from the 1920s. A situation cannot be reproduced. This can be understood on the basis of the experience of world in praxis. In praxis, every situation is potentially new. This is so because, in being with one another, the possibility of a new beginning is always there.

However, in his analysis of boredom, Heidegger at first employs the term 'situation' in a non-terminological sense. In the case of the first form of boredom, namely, waiting for a train at a station, this is boring because the situation leaves us empty. This 'being left empty' (*Leergelassenheit*) assumes that something is expected in this situation (being at a station) which is not fulfilled. (cf. Heidegger, 1995; pp. 104, 106; GA 29/30, pp. 157, 160) In the second form of boredom, that of 'being bored together with …' (*Sichlangweilen bei …*), we are still under the domination of the 'they'. The whole situation – Heidegger's example is that of being at a party during the evening – is both boring and a way of passing the time. Such boredom, which in the everyday is never far from the surface, comes to appearance in the conventions of a social situation. We take time to go to the party, take off those hours, because we want to treat ourselves, we want to take a break from our daily responsibilities and forget about our yesterdays and our tomorrows. We want to simply enjoy the present and forget the past and the future. (Heidegger, 1995, p. 124; GA 29/30, p. 187) In that sense the now of the evening should not pass, it should offer us no possibility for transition (*Übergang*). Time should 'remain standing'. Everyday chronology is a chain of transitions; time is a transition from one 'not yet' to one 'not any longer'. These transitions occur only in a chronological order. In this order, there is on the one hand, no standing time; on the other hand, there is no free-time. In such times, we are, or so it seems, freed from time. But that is not the case; rather, we take for ourselves a time that is free. We bring this time to a standstill. And in this standing time, Dasein is brought to the impossibility of temporal passing. (Heidegger, 1995, pp. 124–5; GA 29/30, pp. 187–8)

It is the passing of time, however, which first makes the situational possible. It is that which always anew appears to us as new. This possibility comes to appearance in the third form of boredom, which Heidegger characterizes with the phrase: 'it is boring for one' [*es ist einem langweilig*] In this case, the time of boredom cannot be calculated. Such boredom can engulf us when we least expect it; perhaps, even when we expect the opposite. (Heidegger, 1995, p. 135; GA 29/30, p. 203) '*It* is boring for one' means that it is indeterminate *what* bores; 'it is boring for *one*' means that the one who is bored is not I as I, you as you, or us as us, but an undifferentiated nobody. Dasein becomes indifferent; it appears not as itself and remains indeterminate: I am as indifferent to the other as she is to me and as indifferent to myself as either.

We can now return to the situation of friendship. The other as friend withdraws from me. This has its roots in the seeking for fulfilment. We saw this seeking in the first form of boredom, in the form of the everyday fulfilments of a one-self. Such fulfilment is manifest also in the second form of boredom, precisely to the extent to which it emerges as Dasein flees the fulfilment of its everyday responsibilities. In the third form of boredom, the situation is more complex. Precisely because fulfilment can emerge in any situation, Heidegger does not give any examples of this third form of boredom.[40] If we compare this analysis to that of Angst, with which it shares many structural commonalities, we notice that Heidegger does not speak of the failure of handy things at all, and only briefly of the failure of the objectively present. Rather, he speaks neutrally of the indifference of entities. If we read this terminological change with the emphasis on mood as the atmosphere of being with one another, then we have an indication that Heidegger has in mind the failure within the domain of praxis, not poiesis. In that case, the fulfilment which is at issue in the third form of boredom is that which is sought in human praxis with others.

We seek fulfilment in the other as friend in the sense of completion. But this seeking encounters the alterity of the other. I cannot become the other. My finitude consists in the fact that my time is always mine. That does not mean that the time which emerges between us and which can always emerge anew between us is an illusion. But it does indicate that this time can only emerge momentarily and that in my being-with I am always again thrown back on my own responsibility – a responsibility which can only be understood from the between-time and between-space in which claims are made on me.[41] Heidegger characterizes the ambiguity of this situation as a 'being forced (*Gezwungensein*) to listen (*Hören*)', a being forced which goes together with the 'most inner freedom' of Dasein. (cf. Heidegger, 1995, p. 136; GA 29/30, p. 205)

To listen to the friend is to hear that which comes from beyond Dasein and to which it responds through its own freedom. Heidegger's analysis here is reminiscent of the sections on conscience in *Being and Time*.[42] The 'voice of the friend' is taken up in this analysis in which Heidegger speaks of an other 'which is with me in the world'. What we truly hear, that which concerns us, wakens us. When we are so awoken, then our affective state is transformed. Hearing is essentially related to the waking of a mood. Waking is always in relation to another, or at least to that which is other. I cannot wake myself; I can at most let myself be awoken. (cf. Held, 1991, p. 40) The mood of fundamental boredom is awoken when Dasein is brought to its own ground. (cf. Heidegger, 1995, p. 131; GA 29/30, p. 198). There Dasein is forced to listen to the other. The order of hearing, the harmony of which we spoke above, which makes friendship possible, no longer exists. I listen to my friend, but he has become alien to me. I experience his speaking only as a veil behind which the friend hides himself and withdraws. The binding harmony between us, which I characterized as timing, is torn asunder. I attempt to find the friend again, but the mood of boredom which has suddenly engulfed me has withdrawn him from me. In this situation something has failed me; to understand what that is we need to turn to the second theme, the power of time.

2) The concept of power is hardly mentioned in *Being and Time*.[43] Surprisingly though, in the lecture course of 1929–30, Heidegger speaks of the 'power of time'. The explanation of this may be that in the mood of boredom Heidegger comes across an experience of time, not as making possible, but as affliction. Michel Theunissen calls this experience that of time as domination (*Herrschaft der Zeit*). (cf. Theunissen, 1991, pp. 37–86) According to Theunissen, if the hypothesis of time as domination is true, then the Kantian thesis of the subjectivization of time is false. (cf. Theunissen, 1991, p. 42) If time is experienced as domination or power, then it is not constitutive of the self, but rather a power over the self. Theunissen explains subjectivization in terms of Schopenhauer's formula: 'before Kant we were in time, after Kant time is in us'. (cited in Theunissen, 1991, p. 39) Hand in hand with this thesis goes a second one, namely, that there is no time in the singular. There are rather 'times': subjective-immanent and objective-transcendent (Husserl), original and vulgar time (Heidegger). Clearly there is more to time than subjective time – there is the time of clocks, the time of the sun and the stars, the datable time of historical events. Hence, the thesis that time is subjective can only be defended if time

is divided between subjective and objective time. But Theunissen rejects this pluralization of time: there is only one time, the difference between the empty time of nature (pure chronology in our terms) and fulfilled time (kairological time in our terms) is in reality not a difference of time at all. That which fulfils time belongs rather to life and experience (*Leben und Erleben*). (cf. Theunissen, 1991, p. 43)

Heidegger explicitly linked his account of temporality to Kant. Kant, he tells us in a lecture course in the years preceding *Being and Time*, had rightly understood time as the structure of human existence, but had failed to conceive time originally enough. (cf. Heidegger, 2010, pp. 336–7; GA 21, pp. 407–8) But, although, for Heidegger, Dasein is temporal, it experiences its temporality through its moods, and as such *suffers time*, and suffers it as that on which it *depends*. This passive relation to time is not so clearly worked out in *Being and Time*. This is so because the stress in that work is on time as making possible and related to this is the emphasis on Dasein as individuated. If, however, we read that work from the perspective of the 1929–30 lecture course, then it becomes clear that the experience of dependence is fundamental. The ecstases of time are not under Dasein's control. Each ecstasis is experienced in mood as overpowering and as breeching the continuity of 'and so on'.[44] The power of time and the temporality of Dasein far from being in conflict mutually condition each other: Dasein experiences time as powerful due to its own existence as temporality. This power of time, however, is not experienced by Dasein continually but rather in the 'between' of rupture. This 'between' is experienced in mood, and in mood we can see time as both a power *over* Dasein and as always coming *from* Dasein.[45]

Theunissen's suggested way out of the domination of time, namely, the tearing oneself from time (Theunissen, 1991, p. 57), is questionable. If we understand temporality on the basis of praxis, the suffering of time shows itself to be the flip side of the joy of acting with time.[46] This joy *with* time – as opposed to joy *against* time – is not recognized by Theunissen; without reference to it, though, we cannot understand the time of friendship. The power of time is such that it can sometimes be an affliction and sometimes leave us in peace. (cf. Heidegger, 1995, p. 98; GA 29/30, p. 148) According to Heidegger, one is held in by a hesitating time in being bored. (cf. Heidegger, 1995, pp. 99–100; GA 29/30, p. 150) Time shows its power precisely then when it fails us in withdrawing. In the first form of boredom, it fails us when things do not occur in their time. To expect something to occur 'in time' is to base our expectations on the past, on past regularities. The power of time is the basis of the regularity of things: we encounter things at particular times (and in particular places). Everyday experience is based on this temporal regularity and consistency. Such regularity is expected into the future too. We live into the future on the basis that past regularities will be maintained. In such cases as these, expectations are disappointed and we have to wait; time becomes 'long': the German term for boredom – *Langeweile* – means literally a 'long while'. What is only five minutes is experienced as lasting forever: therein we encounter a non-chronological temporality.

Not only things fail us; we find ourselves bored in friendship also. This is a deep boredom in which the power of time discloses itself as 'mysterious' [*Rätselhafte*] and 'hidden'. (cf. Heidegger, 1995, p. 147; GA 29/30, p. 221) The mysteriousness of time consists in the fact that it is both a happening of openness and of the indifference, i.e. the hiddenness, of entities in their totality. The logic of the failure of a community of friends and the reciprocal relation of openness and hiddenness are rooted in this mystery. These two logics are, in truth, one and the same. As Heidegger says, all failure (*Versagen*) is a saying (*Sagen*) that is a making manifest (*Offenbarmachen*).[47] The mystery of time can be articulated in this way: time makes possible both the being with of friendship as also the withdrawal of the other into indifference. This ambiguity is most clear when Heidegger states: 'time, which Dasein itself in its totality is, binds Dasein'. (Heidegger, 1995, p. 147; GA 29/30, p. 221) It seems as if Dasein is bound in itself. What is at issue here is a law that Dasein does *not* give itself: Dasein is at once enchained and made possible as free by time. As such, temporality constitutes a Dasein, which, however, is not autonomous in its relation to time. Hence, while Dasein is temporal, it does not subjectivize time, as it is not a subject to begin with. Dasein's experience of itself is shaped by its temporality: an experience of itself as made possible and freed up by that which engulfs and enraptures it – in terms of its affective states, this is manifest in boredom and Angst, on the one hand, and in joy, on the other. These moods correlate to the slipping away and the emerging of time, respectively.

Deep boredom arises not through a failure in Dasein's projections, but rather in the failure of the 'coming' to entice. In deep boredom, Dasein does not 'expect anything from entities as a whole in any respect, because there is not even anything *enticing* about entities anymore' (Heidegger, 1995, p. 147; GA 29/30, p. 221 [my emphasis]) Here, the future in which entities entice Dasein fails. That which entices in the future is not Dasein itself – not Dasein in its own projections – but rather always an other. The enticements of the future withdraw and fail Dasein, when the other's future fails it.

What is experienced in this failure is the dependence not on time as chronology, but rather on a time that cannot be measured. It is this time which 'binds'. When the other no longer entices, time as essentially enticing appears precisely as its absence. It is quite consistent of Heidegger to stress the double role of time as failing and making possible. In the failure of an order, a deeper order is revealed: this is the ordering principle of *Being and Time*. The deeper order is a 'repetition' of the failed one. On this basis, Heidegger can say that 'the temporal entrancements can be ruptured only through time itself'. (cf. Heidegger, 1995, p. 151; GA 29/30, p. 226) But in this case the failure of an order does not reveal an order that was always already there, but rather what is disclosed to Dasein in this failure is the situation of 'essential action'. (Heidegger, 1995, p. 151; GA 29/30, p. 227) In this failure, the moment of vision is revealed for the first time as the moment of essential action, in which a new order is suddenly possible. This moment of vision is the 'vision of resolute disclosedness for action'. (Heidegger, 1995, p. 151; GA 29/30, p. 226) This moment of vision comes over Dasein as a strange and alien power. Dasein experiences here its dependence on time, and it is precisely this dependence which announces itself in the failure of friendship.

In the moment of failure:

> the whole expanse of the entire time of Dasein is there and not at all specifically articulated or deliminated according to past and future. Neither merely the present nor merely the past nor merely the future, nor indeed all these reckoned together – but rather their *unarticulated unity* in the simplicity of this their horizon all at once.' (Heidegger, 1995, p. 148; GA 29/30, p. 222 [emphasis in original])

It is precisely this unity which for Heidegger lies at the basis of resoluteness and hence the possibility of action. Mood and action are brought into harmony here in a manner which was not the case in *Being and Time*. Heidegger speaks here of 'essential action' without hesitation. While in *Being and Time*, 'action' is used only with caution (cf. BT, p. 276; H, p. 300), here it is used freely. This is due to the central role of transformation (*Verwandlung*) in the 1929–30 course. This emphasis is closely connected to the development of the understanding of historicity in this lecture course. In the moment of essential action, there is the possibility of transformation. Time is not brought to a stand, but rather comes to appearance as transition (*Übergang*). The possibility of Dasein is the possibility to contribute to this transition. In this possibility, the kairos is opened up as demise and coming forth, as beginning (*Anfang*). That which emerges here is a *new* order. The failure of the order of praxis demands a response from Dasein. Dasein is faced with its own responsibility. In resoluteness, it responds to this challenge in the form of 'essential action'. Resoluteness is thus understood as a response to a failure, a response which is dependent on temporality. What becomes clear here is that which remains obscure in *Being and Time*, namely, that temporality is experienced by Dasein as a power which is both over it and in it.

This opening up of a new praxical order is already political. The between-time of being-with, that time most clearly evident in public form in friendship, opens up a political space, which however is subject to transformations in the affective state. What this suggests is that affectivity in Heidegger – specifically fundamental moods – needs to be understood inter-subjectively, and that the power of time understood inter-subjectively is a power of transformation that transcends all attempts to ground it. The time of praxis is a groundless, abysmal time. To the extent to which time reveals itself as power, to the extent to which it fails and withdraws from us, time is the power of emergence as well as decline. A radically kairological understanding of the time of such emergence and decline can only be based on an ontological uncertainty regarding Dasein, the public world and historical change itself. Heidegger, in the wake of *Being and Time*, thematized this uncertainty as freedom. It is to the question of freedom that we must now turn.

3

Freedom, Contingency, Truth: Time as Emerging

The order of time is a split order. To understand what this means it is necessary to clarify what has been already referred to as the emergence of time. The difference of kairos and chronos is one with respect to time as emergence. The emergence of time is a phenomenon of freedom. This is so because emergence constitutes a rupture with passing, as that which grounds causality as the relating of events to a past cause. Understood as a phenomenon of beginning, time as emergence means a break with the past. In other words, such emergence is contingent. A causal explanation is one which indicates a relation to the past, in which the explanation for an event can in principle be exhaustively gathered from the past. That which emerges contingently could always have been otherwise, and although as that which *has emerged* can be causally explained, *in its contingency* it escapes such explanation.[1]

In the following chapter it will be shown to what extent it can be meaningfully said that 'time emerges'. At first the sentence may appear to contain a simple confusion. It may be objected that while things may emerge *in time*, time itself emerged if at all only once at the beginning of the universe. Nevertheless, the idea that time does not just begin once, but is a continual possibility of beginning, is one which is contained in the concept of continual creation, which we find in different forms in such thinkers as Maimonides, Descartes and Malebranche, amongst others. The insight which underlies this thought is that of the uncertainty of the future and the limits of any causal explanation. This uncertainty is understood in this concept as dependence on God. Descartes' argument for continual creation is based on the thesis that infinity is more perfect that finitude, something Heidegger fundamentally denied.[2] In consequence, for Heidegger the emergence of time is without ground. This can only be understood on the basis of a radically anti-Cartesian, or non-causal, understanding of freedom.

I Contingency and freedom

Heidegger defines freedom in *Being and Time* as follows: 'freedom *is* only in the choice of the one [possibility], that is, in bearing the fact of not having chosen and not being able also to choose the others.' (BT, p. 263; H, p. 285) The question has long been debated as to whether what Heidegger describes is really a matter of contingency or must we rather assume that whatever is possible is either already actualized or will be actualized. (cf. Vuillemin, 1984) The underlying question here concerns the contingency of the future. If the having-been is 'awoken' by the future, then a further question concerns the contingency of the having-been. If kairological time means a sudden transformation – and if it does not just appear to be such on the basis of a lack of information or knowledge – then such contingency must be accounted for ontologically. If the kairos means a transformation into the future, then it is a moment 'between' the having-been and the future. In this sense, the kairos is the time between liberation from … and a freedom for … between negative and positive freedom. If the kairos is, then this 'between-space' is purely contingent.

I.1 The having-been and the past: Contingency

To this point, the futuricity of the kairos has been stressed. But what happens suddenly concerns Dasein not in its future, but in its present. Dasein must deal with it 'now'. But it seems that even to speak of the present here is mistaken. That which happens suddenly, we say, is over before we know what happened. In this sense, human action always happens subsequently. The event is already past, we have to deal with its 'trace'. Hence, the question is, how Dasein has access to this 'having-been' event.

For Heidegger, Dasein's access to the past is something different to the access to the having-been. For one thing, the past cannot be changed, the past is 'unchangeable, closed off, and never to be brought back … The past lies before the door of the present and can never go back and into it.' (Heidegger, 1989, p. 108) Every point in time which has passed is, in this sense, past. In the chronological, there is no repetition: what was is no longer, and for this reason cannot be changed; we have no *direct* access to it anymore. Strictly speaking, if time could only be understood chronologically then every point in time would fall into the past and could never return.

Aristotle had this characteristic of the past in sight when he spoke of 'conditional necessity'. According to Aristotle, the past and the present are necessary, while the future is in a certain sense contingent. The necessity which is in question here is a *conditional* necessity. Aristotle defined conditional necessity in the following way: 'what is, necessarily is, when it is; and what is not, necessarily is not, when it is not.' (Aristotle, *On Interpretation*, 19a23) *If* something is, *then* there is no possibility to deny its existence. This is not to say that it exists *out of* necessity. It was not necessary that what occurred did occur. But once it does occur, it cannot be undone. A concept of causality lies at the basis of the past so understood. Nothing can have efficacy on the past. In the domain of human action, we cannot always predict what effects our actions will have, but once these effects have taken hold, we cannot go behind them and change the past causes. The necessity here lies in the fact that cause and effect have *actually* happened.

What is actual is objectively present. That which is past is no longer objectively present. The past is a mode – the mode of the no longer – of the objectively present. In this sense, Dasein is never authentically past. As Heidegger puts it, '"As long as" Dasein factically exists, it is never past.' (BT, p. 301; H, p. 328) In its relations to entities Dasein is its possibility. As with past entities, Dasein does not encounter its actuality directly, but rather as the possibility to relate to itself as actuality. As such Dasein is never actual in the sense of an objectively present thing – it has simply the possibility of understanding itself as objectively present – and as such it is never 'past', but rather having-been (*gewesen*). Dasein's 'past' 'essences (*west*)' as possibility. Past things also are not immediately accessible as past, but rather only in relation to the having-been, that is, to the possibility of relation to what has been.

In his lecture course from summer semester 1927, Heidegger states: 'Dasein is also in a certain way in time, for we can view it in a certain respect as objectively present.' (Heidegger, 1988, p. 271; GA 24, p. 384 [translation modified]) We have the possibility to view Dasein, both in ourselves and others, as objectively present. This is consistent: the point of Heidegger's account of authenticity is that Dasein for the most part views itself as objectively present – hence views itself inauthentically – and must first *gain* its authentic self. It follows from this that Dasein can also be past. Heidegger appears to confirm this conclusion when he says that Dasein can sensibly speak of past worlds. (cf. BT, p. 348; H, p. 380) But it is precisely when Dasein wishes to take responsibility for its past that it no longer views the past as objectively present, but rather as a possibility that makes a claim upon it. Heidegger characterizes this experience as rejoinder (*Erwiderung*), that is, as a form of response.

In Chapter 1, it was discussed how Dasein relates to its past through repetition, and the extent to which repetition is a rejoinder. Dasein must make a rejoinder on that which makes a claim on it. The past, which Dasein encounters only as an object of consideration, can never be *its* past in an emphatic sense. This is so because when the past makes a claim on Dasein and when it must take responsibility for that past, then that past is no longer an object of consideration but rather 'happens' (*geschieht*) in its action, precisely its action as repetition.

'The having-been arises out of the future' [*Die Gewesenheit entspringt der Zukunft*] (cf. BT, p.300; H, p. 326) The future for Aristotle is the domain of practice (both praxis and poiesis), because it is only sensible to deliberate about the future. (cf. Aristotle, *On Interpretation*, 19a7–22 and *Nicomachean Ethics*., 1139a13f.) We deliberate, namely, only about that which could be different; the past cannot be other than it is. Theory takes the past, the unchangeable, as its model, while practice is orientated towards the future, the domain of the possible. If the future is to be understood in continuity with the past, it remains unclear how we are to account for the accidental. If the future is contingent, then it must be possible for this continuity to be broken or ruptured.[3] Contingency, in other words, cannot just be a matter of the future, it must also happen in the present. The present, though, for Heidegger is released out of the future past. If the present is contingent, its contingency is by the temporalizing of the future.

What is at issue here is fate (*Schicksal*). It is a matter of fate – to employ an example of Aristotle's – whether a sea battle will happen tomorrow or not. For this reason, the question cannot be decided: there is here an exception to the principle of the excluded middle. (cf. Aristotle, *On Interpretation*, 19a30–3) Destiny does not simply affect the future, however. The having-beenness of Dasein is conditioned by its throwness. Indeed, Heidegger introduces destiny in terms of throwness. As he says:

> The finitude of existence thus seized upon tears one back out of endless multiplicity of possibilities offering themselves nearest by – those of comfort, shirking and taking things easy – and brings Dasein to the simplicity of its *fate*. This is how we designate the primordial occurrence of Dasein that lies in authentic resoluteness in which it *hands itself down* to itself, free for death, in a possibility that it inherited and yet has chosen. (BT, p. 351; H, p. 384 [emphasis in original])

Dasein comes back to itself as possibility out of the plurality of possibilities offered to it. Heidegger understands this possibility as the 'simplicity of fate'. In this way, he restated what he had already thematized as thrownness: The possibility of the past, for which Dasein is itself not responsible, but to which it is delivered over. In the same way, fate like thrownness can only be taken over appropriately in the mode of resoluteness. But what Heidegger characterizes as fateful is the historicity of thrown Dasein. The possibility of Daesin is not made by it. It is also not the expression of a transcendental ego. It is rather possibility which it inherited, which was transmitted through tradition.

If the above interpretation is correct, then the sentence 'yesterday there was a sea-battle' is just as problematic as the sentence 'there will be a sea-battle tomorrow' because the having-been arises out of the future and is as such contingent. What actually happened cannot be changed; but the having-been is the possibility of the relations of Dasein to the past, which is first brought to light through repetition. (cf. Figal, 1988, pp. 320–1) That which is transmitted through repetition is so, according to Heidegger, 'only if we grant the possibility of transformation.' (Heidegger, 1984, p. 155; GA 26, p. 197) The actuality of the past can only be recognized as a moment of this repetition. It is only accessible through the possibility of the relations to it.

What is apparent here is that contingency does not arise out of the futuricity of Dasein so much as out of the latter's character as possibility. As such, Heidegger's thesis of the priority of the future – no matter how much this has been stressed in the last two chapters – needs to be qualified. It is indeed the case that we can see the character of possibility of Dasein most clearly in the temporal ecstasis of the future; but this character arises out of the temporality, not out of the futuricity, of Dasein. If this were not so, it would not be understandable that Dasein could take over its having-been as possibility.

Every past point in time can be said to be unique in the sense that it cannot be repeated. The hero, whom Heidegger says Dasein chooses in repetition (cf. BT, p. 352; H, p. 385), is presumably characteristically unique. In a lecture course from winter semester 1931–2, *On the Essence of Truth*, Heidegger states that 'history is … always a matter of the unique task posed by fate in a determinate situation of action'. (Heidegger, 2002b, p. 66; GA 34, p. 91 [translation modified]) History is as such unique. Repetition forms the access to history in its uniqueness. The unique allows itself then to be repeated. In his first lecture course on Hölderlin, Heidegger states, 'Unique means … precisely not once objectively present and then past, but rather having-been and therefore in the constant possibility of essential transformed unfolding'. (Heidegger, 1989, pp. 144–5) In other words, that which is unique in history is precisely not that which can be uniquely dated, but rather what does not appear inscribed in the historical record. The historical unique is not that which is documented and carefully preserved for posterity, but rather that which escapes such record either in its self-evidency or in its untimeliness. 'Repetition' then means the re-occurrence of the past world in its self-evidency and untimeliness in such a manner that it becomes once more alive. To return to the hero: it is not that which can be made the object of reports of the hero's actions that are 'repeated', but rather his manner of relating to the world, which is manifest – but only implicitly – in his actions.

What occurs through such repetition is a transformation of the past itself. That which was familiar – the past as related in the transmitted accounts – becomes strange in a remarkable way through its repetition. The possibility of dealings with and in a world can only appear strange if experienced as inappropriable. (cf. Held, 1991, pp. 31–3) The more, for example, we (as Heidegger puts it later in the Anaximander fragment) 'insist on thinking Greek thought in a Greek manner' (Heidegger, 2002, p. 253; Hw, pp. 332–3), the more its world becomes yet stranger, and what thereby comes to appearance is the 'constant possibility of essential transformed unfolding'. (Heidegger, 1989, pp. 290–3)[4] Heidegger terms this strangeness of the past the 'unsaid'.

For Heidegger, this 'unsaid' is 'being'. One might want to object here that what is at issue is not so much being as the everyday, indeed the most everyday of the everyday, that which is most familiar. But that which is closest to Dasein is not the most familiar to it. What is familiar is rather Dasein's being-in. (cf. BT, pp. 50–1; H, p. 54) 'Being-in' is being familiar with something, dwelling with it. 'In' does not mean a relation to the objectively present, but rather the familiarity with the world which first makes all relations possible. Familiarity is not with an entity or entities, but rather indicates the possibility, the making possible, of familiarity with innerworldly entities. Dasein's access to being is precisely through this familiarity. It is for this reason that Heidegger begins the analyses of *Being and Time* with a discussion of handiness, what he considers the most familiar manner of dealings with things in the world. In the everyday, in the most everyday of the everyday, the inner possibility of a world opens itself up. Repetition is no arbitrary choice; it is rather an estrangement from the having-been world and the everyday world of the present. This estrangement is, in fact, an estrangement from chronological time. This is so because the past as a continual sequence of nows is interrupted; familiarity rooted in this continuity of the past is destroyed through an estrangement in which the past world appears strange and having-been.

Such a disclosure of the strangeness of an other, past world, has an effect on the present familiar world. This is so because the strangeness of the other world is such precisely in making a claim on Dasein. That which appears strange to Dasein makes a claim on it by showing it the possibility of being otherwise.

Dasein is delivered over to this possibility. This claim of the stranger – the having-been Dasein in a past world – challenges Dasein to see its world as strange – though not to see it from the standpoint of the stranger, which is closed off. (cf. Waldenfels, 1990, p. 64) What becomes apparent here, rather, is the making possible of history in the sense of uniqueness. It is then when we seek to repeat a having-been world, and as such to be authentically historical, that the strange world and the own world appear in their strangeness, that is, in their uniqueness.[5] In this moment, Dasein is thrown back on the possibility of its world. Repetition does not bring Dasein, as Heidegger in *Being and Time* still thought, simply to the original experience of itself (cf. BT, pp. 19–20; H, p. 22), but rather to the uniqueness of its own historical situation.

It is admittedly hard to think how we can express the truth of the having-been in this situation. What is the nature of the temporal and historical situation in which we can speak of the truth of the having-been? Aristotle assumes that this truth can be expressed in a present in which there is an asymmetry between past and future. Assuming such an asymmetry, past truths are in principle unproblematic because the past can no longer be changed. If, however, we follow Heidegger's account of repetition, we can no longer assume such an asymmetry and the contingency of future truth becomes not a distinctive feature of the future alone, but rather characteristic of Dasein's temporal being as an entity characterized by possibility. This thought is expressed by Heidegger when he states, '[G]enuine historical return is the decisive beginning of authentic futuricity.' (Heidegger, 2002b, p. 7; GA 34, p. 10) This reverse movement (*Ruckgang*) opens up the character of Dasein as possibility.

It remains the case that Dasein has a chronology. It cannot change where it was born or what it has done. In the repetition of its biographical or historical having-been it relates to the actual, which can neither be wished away nor – as past – be repeated. The actual cannot be repeated, because it is chronologically fixed. Repetition, however, is a setting in relation with the actual past. The past can never be set aside, because Dasein exists factically. Through its thrownness, Dasein finds itself in a particular people with a particular past. This past *ought* not remain past, however. Dasein's obligation is to take responsibility for the factically distinct possibilities in which it is thrown. These possibilities remain past so long as Dasein does not make a rejoinder to them. Such a rejoinder assumes that the past has made some claim on it. Dasein wishes, on the one hand, to free itself from the past, but on the other, to allow a reoccurrence of the past and in so doing to take responsibility for it.[6] But, to the extent to which the past is understood as a continuity and as governed by a causal chronology, the encounter with it in repetition is also a break with causality. Through this break, the actual past is repeated and transformed into a having-been possibility.

Aristotle's account of the truth of the future assumes both the relations of causality and the arrow of time. According to both of these, effects do not work backwards. If, however, the past is as much a realm of possibility as the future, then causality and the future directedness of time are both called into question.

I.2 Freedom and causality

According to Hannah Arendt, there is a hiatus between liberation and being free. (Arendt, 1978, p. 204) Freedom does not follow automatically from liberation; there is rather a breach which must be traversed, in order to reach the new freedom. The possibility of this breach, this rupture is for Arendt that which makes revolution possible. It is possible in historical time for a chasm to open up, in which the possibility of the new emerges, in which, however, there is no assurance as to whether that future will bring good fortune or catastrophe. In this 'between-space', the contingency of the temporality of Dasein comes clearly to appearance. The kairos is the order of time of this between-space between liberation and freedom.[7] In his 1929 essay, 'On the Essence of Grounds', Heidegger states, 'Freedom as transcendence … is not only a unique "kind" of ground, but the *origin of ground in general. Freedom is freedom for ground.*' (Heidegger, 1998, p. 127; Wm, p. 162) Since St Augustine, at least, such a freedom to ground has been understood as will. (cf. Arendt, 1978, pp. 82–107) Will is, according to *Being and Time*, rooted in care. Heidegger alludes to this when in 'On the Essence of Grounds' he understands 'willing' in the sense of the 'for-the-sake-of-itself' (*Um-willen-seiner*) (Heidegger, 1998, p. 126; Wm, p. 161) Dasein surpasses itself in the 'for-the-sake-of'. This means that Dasein can have a world only because of the for-the-sake-of. The will, of which Heidegger is speaking here, is that which makes all relations and engagements possible. As a surpassing, it forms the 'for-the-sake-of'. (cf. Heidegger, 1998, p. 126; Wm, p. 161) Heidegger terms this surpassing freedom. It would at first sight appear that Heidegger is here, in keeping with the tradition of the Stoics and Augustine, grounding freedom in the will. (cf. Arendt, 1994, pp. 219–20 and 1979, pp. 71–107) But in reality he is claiming the reverse: the human being is not free because of his will, but has a will because he is free. Will is rooted in care and care is nothing other than 'being free *for* its onwmost potentiality-for-being, and thus for the possibility of authenticity and inauthenticity'. (BT, pp. 179, 191)

As Heidegger understands freedom in that work, it is not determined by will, but is rather to be understood as *response*.

To be free is to be responsible: a free being is one which is responsible for its actions. For Heidegger, responsibility presupposes being delivered over (*Überantwortung*). Dasein can bear responsibility because it is must respond to its thrownenss – its being delivered over. Heidegger makes this clearer when he speaks of anxiety: '*Angst* brings Dasein *before its being free for* … (*propensio in*), the authenticity of its being as possibility, which it always already is.' (BT, p. 176; H, p. 188) If this is so, and if the will is grounded in freedom, then the will has the structure of response. Transcendence, then, is not an arbitrary stepping over of entities, but rather a response to being in them.[8] The consequence of this is that Dasein cannot be its own origin. Freedom cannot mean originating from itself (spontaneity).

In the introduction to his lecture course on Hölderlin in the winter semester of 1934–5, Heidegger distinguished between beginning (*Anfang*) and starting (*Beginn*): 'A start is the onset of something; a beginning is that, out of which something arises or springs forth [*entspringt*] … The start is immediately left behind, it vanishes in the progress of an occurrence. The beginning, the origin [*Ursprung*], on the contrary comes first to appearance in the occurrence and is fully there only at its ending.' (Heidegger, 1989, p. 3; cf. McNeill, 2006, pp. 116-19) The human being can start; only a god can begin. The beginning is as such outside all chronology; the human being can deal with its traces, never with it as such.

This seems to contradict the stress I placed in Chapter 1 on the ability to begin (*Anfangskönnen*). In that chapter, I argued that the human being is kairological to the extent to which it is capable of beginning, and that this was understood as to do with Dasein's natality. Natality was understood in Chapter 2 as to do with praxis not poiesis. There seems here to be a twofold shift in Heidegger's thinking: on the one hand, the possibility of originating as a human capacity is excluded; on the other hand, freedom is understood as the freedom to ground and hence as poietic.

The second issue here – that of the poietic – will not be addressed until the next chapter. With respect to the first issue – that of natality and beginning – what we find is a change of emphasis, which becomes manifest in the lecture courses in the 1930s, first with that on Hölderlin. Natality as possibility to begin characterizes an entity which can encounter origins. An origin is the opening up of a world, over which the one born has no control, but which it experiences as an overwhelming power. The ability to begin as natality is the possibility of responding to the event of beginning. When Heidegger distinguishes sharply between beginning and starting, and does so in terms of a chasm between gods and humans, then what is of concern is not so much birth – as the possible entry of the human into history – but the possibility of history itself. History is only possible when there is a rupture with the past. This rupture is the condition for the entry of human beings into history. Freedom is not made possible through this entry, but rather the reverse. Freedom, then, responds to a beginning in which history is first possible. It is responding to the unconcealment of being in the kairological situation of action.

At the same time, the freedom to ground is a matter of beginnings. When I stand up from my seat – to use Kant's example from the third antinomy (Kant, *Critique of Pure Reason*, A 450; B 478) – this act presupposes causal relations with the past. If the kairos is not simply an epiphenomenon of chronos, then a new beginning needs to be thought otherwise than this. Furthermore, if freedom is not a form of causality, but causality a form of freedom (cf. Heidegger, 2002a, pp. 20–1; GA 31, pp. 27–8), then the temporality of Dasein must make possible for it to withdraw itself from the temporal sequence and in doing so not to enter into timelessness, but rather to experience (kairological) time in its full *intensity*. To show this, we must first discuss why it is that chronology must be understood in causal terms and how, through a certain intensity of time, a rupture in causality appears, and how, finally, this free rupture relates to causality.

It is important at the outset to be clear that this question does not have to do with the issue much debated in analytic philosophy of the difference between cause and motivation. Also the question of causal analysis in the human sciences is not a matter for discussion here. (cf. respectively Davidson, 1980 and Ricoeur, 1984, pp. 182–90) Causality is not being understood here on a mechanical model, but rather most generally as the relation between two objectively present entities in which a change in one is caused by an act of the other. (cf. Heidegger, 2002a, pp. 133–4; GA 31, pp. 190–2) The question is now whether causality so understood assumes chronological time.

When something is a cause, then that which it effects must be present. Furthermore, the cause and that which is caused must come into contact with one another at the same point in time. The movement can only go in *one* temporal direction. Certainly that which is affected can have a reverse effect on the cause, but this is not an effect on the past. Causality assumes a forward directionality of time. In the realm of human action, when we are concerned with the justification of action, we look to the past for the 'causes' of an action (cause in the neutral sense of that which brought it about). In this sense, causality assumes chronology as the directionality of time. Similarly, chronology assumes causality. Chronology is without gaps: 'the succession of nows is uninterrupted and has no gaps'. (BT, p. 388; H, p. 423) There is no now point without a previous and a following now point, in the same way as nothing is without cause. If there were no causality, we could not talk of time: if there was no necessary connection between before and after there would be no time, as there would be no reason for one now point to pass into another. Such a passing is movement and, for Aristotle and the philosophical tradition following him, there is no movement without reason or cause.

In the Aristotelian tradition, a movement of time without ground is nonsensical. Indeed, it cannot be understood at all on the basis of a chronological account of time. To make sense of it, we need to think through the phenomenon of the intensity of time.[9] Time has an intensity in the sense that its movement is no longer dispersed but, rather, forms a unity. In this time, Dasein is taken back out of the lostness in concern and placed within the power of time. Time opens itself up as a power, which is enticing, demanding, even seductive.[10] Through this, we find ourselves 'momentarily' in the transition to a new order. Heidegger understands this phenomenon of intensity as occurrence (*Geschehen*). Already with reference to *Being and Time* we have argued that occurrence expresses a discontinuity of historical time. Understood in terms of causality, an occurrence would mean a coming to be of a beginning, which neither follows from nor through what went before, but is rather a beginning 'according to time'. (Heidegger, 2002a, p. 152; GA 31, p. 220)

Chronologically understood, such an origin would be impossible; but the occurrence is a movement of transcendence in which a world first opens up. Such a movement has the character of an interruption.[11] An interruption is not an event in the sense of an occurrence *in* time. It is rather a revelation in the moment of vision, which cannot be explained on the basis of what went before. In such an interruption, a liberation from the previous order and indeed from chronological time itself happens.[12] The occurrence is not simply a liberation, but is at the same time a making manifest and as such a disclosure of a situation for action.

Heidegger stresses that the temporal sequence, as Kant presents it, 'does not mean just one thing after another in order of their appearance and disappearance, but a unidirectional, irreversible succession'. (Heidegger, 2002a, p. 136; GA 31, p. 195) Decisive in the understanding of temporal sequence is 'the uniquely directed order in the presence of one [cause] and the other [effect]' (Heidegger, 2002a, p. 136; GA 31, p. 195) This understanding of temporal order is based on an account of objectively present entities. As such, this account assumes the position of an unengaged observer. Heidegger points out that in speaking of action Kant does not do so from the position of the agent, but rather views action as objectively present. This is so because for Kant action is a name for efficacy, which is directed towards objectively present things. (cf. Heidegger, 2002a, p. 136; GA 31, pp. 196–7) On the basis of this order of time of the observer, Kant derives both permanence and continuity from change in the analogies of experience. This is based, however, on an ontology which places causality in the centre and – according to Heidegger – does not accord freedom its proper metaphysical place. (cf. Heidegger, 2002a p. 168; GA 31, p. 246)

How then would a kairological order appear in this context? Kant's temporal order is rooted in Aristotle's understanding of the conditioned necessity of the past. Past time is behind us and unchangeable. An understanding of the kairological understanding of time must then transcend both the Aristotelian and Kantian accounts of time and causality. Heidegger attempts to do this in his discussion of occurrence (*Geschehen*). He denies that occurrence is 'a succession of processes, a changing appearance and disappearance of events' (BT, p. 347; H, 379) The occurrence is rather fateful, that is, the disclosure of a situation of action. It is, however, for Heidegger enigmatic how 'this occurrence, as fate, is to constitute the whole "connection" of Dasein from its birth to its death.' (BT, p. 353; H, p. 387) The enigmatic here is how a new order can emerge out of freedom. The movement of the occurrence is enigmatic because it cannot be explained in terms of what was previously there. (cf. BT, p. 355; H, p. 389) This enigma can only be resolved through a rethinking of freedom, which is precisely one of Heidegger's principal concerns in the wake of *Being and Time*.

According to Heidegger, human beings belong to freedom. (Heidegger, 1998, p. 145; Wm, p. 187) This follows consistently from Heidegger's thesis discussed above that the will is grounded in freedom. The human being belongs to freedom because only in freedom does she have access to being: 'Being is the solely and genuinely "in itself"; and hence the originary nature of the understanding-of-being and … of freedom' (Heidegger, 1984, p. 147; GA 26, p. 186) This means that freedom is at the basis of the understanding of being and hence of being itself. In 'On the Essence of Grounds', Heidegger employs 'freedom' and 'transcendence' practically as synonyms, and it is only through transcendence, through the stepping over of entities, that being is. Transcendence forms the being-possible of Dasein. It is only due to its transcending of entities that Dasein has the possibility to understand itself as other than objectively present. Dasein's relation to its being is through understanding, that is, through projection. Its being is not 'there' for it, but rather is that on to which it *projects* its understanding of itself.

Dasein understands itself as its own in its projection on being. Dasein exists only in and through this projection. Dasein's own being is its potentiality-of-being (*Seinskönnen*), which Dasein does not possess as its own, but rather which it *is*. That means that Dasein exists in this difference between its being and itself as an entity. To exist in this way is to be a possibility, that is, to exist *freely* (cf. Heidegger, 1988, p. 276; GA 24, pp. 391–3) If Dasein had its freedom as an attribute, then it would have it as an actuality, but in such a case freedom would be understood as objectively present and as a form of causality. Only as *possibility* can Dasein relate to actuality, that is, through transcending it towards being. 'To be free is to understand oneself from one's own potentiality-of-being'. (Heidegger, 1984, p. 214; GA 26, p. 276) Such self understanding is only possible if and insofar as Dasein projects itself over entities and reflects on itself as beyond that which is objectively present, that is, as possibility. In this context, Heidegger says, 'Since projection unveils without making what is unveiled as such into an object of contemplation, there is present in all understanding an *insight* of Dasein into itself.' (Heidegger, 1988, p. 277; GA 24, p. 393 [emphasis in original]) This insight is the reflection of Dasein on itself in its understanding of being; Dasein's understanding of being constitutes it as an entity which exists in understanding. This insight is not free-flowing knowledge, but arises out of the genuine freedom of Dasein, which constitutes its projection. In temporal terms, this freedom of the understanding of being is to be accounted for on the basis of its character as occurrence.

As already said, the kairos happens before we know *what* has happened. The freedom towards the kairos is a freedom to respond. Past is transformed into possibility in this experience. Such a transformation clearly involves a relation of return to the past,[13] but this is not a step by step return through thought, which follows a causal chain, but rather 'the historical past is not defined through its *position* in the having-been [*im Gewesenen*], but through its future. What is determinative is … *the future in its possibility*.' (Heidegger, 2002a, p. 147; GA 31, p. 213 [translation modified; emphasis in original]) This is not a matter of chronological distance: an event which lies hundreds of years in the past can throw deeper shadows on the present than a more recent event. In this sense, the meaning of the historical past is never fixed with any finality and completeness. An event which has-been is not objectively present; it is accessible only through the traces of it which have been transmitted to us.

These traces lead us, as already discussed, into the strange. This strangeness is not to be explained in terms of its causes or effects; rather, it lies in the possibilities of an other world. These possibilities do not exist in themselves; their uniqueness becomes apparent in repetition. The presupposition of such repetition is that the past is not dominated by causal relations. If causality were the last order, then repetition would not be genuine because it would amount to an attempt to make that which was objectively present in the past present again, in the form of representation. But possibility cannot be represented because it is not an object; rather it discloses itself in action. Being is only 'there' in repetition. The relation to entities presupposes an understanding of being, but this understanding is itself projected, and if thrownness appears in the endless attempt to repeat birth, then understanding is itself repetition.[14] That which only shows itself in repetition, escapes chronology. This is so because chronology binds every event to one or more causes and thereby preserves a chronological unity. The attempt to construe the having-been as past objectively-present events shows the strangeness which only through force can be integrated into chronology, or indeed remains outside it as a madness pushed outside the chronological domain. In this strangeness, a truth announces itself, namely, the disclosure of that which is hidden and distorted in chronology. This truth is precisely a transcending of causality and as such is the truth of a revolutionary time. To understand this, we must examine both the chronological and kairological temporality of truth.

II Truth and time

II.1 The movement of truth and chronological time

Against the traditional understanding of truth as a static relationship between two poles which are both present (*adaequatio rei et intellecti*), Heidegger appeals to the non-theoretical understanding of truth in Aristotle's *Nichomachean Ethics*, book 6. For Heidegger's understanding of truth, it is vital that truth is understood here in the context of dealings with entities. Truth is not here an abstract measure, which can be applied to entities, but is rather that which emerges in dealings with them. Such emergence is at the basis of the Greek word ἀλήθεια As is well known, Heidegger translates this as 'unconcealment'. Truth in this account is an emergence out of concealment. The movement of such emergence does not come from entities themselves: entities are only in the open or in their withdrawal from the open. They are 'wrested from' (*entrissen*) concealment; Heidegger also refers to this wresting as a robbery (*Raub*). (BT, p. 204; H, p. 222)

Truth, according to Heidegger, stands in 'a primordial connection with *being*'. (BT, p. 197; H, p. 213). For this reason, the phenomenon of truth is fundamental to the analyses of the first section of *Being and Time*.[15] Being cannot be brought immediately to appearance, because it is only disclosed through a 'logic of failure'. In this sense, §44 of *Being and Time* is the positive side of Heidegger's analysis. Being is disclosed in the failure of entities. How that happens, however, is the task of § 44, prior to – in section two of the work – 'repeating' (in Heidegger's sense) the analyses of section one.[16] This role of truth is not only dictated by the architectonics of *Being and Time*, but is motivated by the fundamental issue of that work. Dasein does not first seek truth for Heidegger, but rather encounters it in its dealings with entities.[17] Truth only becomes thematic when the entity becomes conspicuous and the relation of Dasein to it becomes problematic. To clarify Heidegger's thesis here, I will relate this account of truth to practice.

According to Heidegger's interpretation of Aristotle, the mode of carrying out of *φρόνησις* is *βουλεύεσθαι* (cf. Heidegger, 2003, p. 99; GA 19, p. 144), which means a consideration, a discussion, of something, which is not present, but must rather be uncovered. (cf. Heidegger, 2003, pp. 99, 102–3; GA 19, pp. 144, 148–9.) Phrōnesis is only possible in the case of such absence. As long as action is caught up in the 'in order to' of the handiness, the agent does not escape from the entity. The movement of uncovering, which is fulfilled in the overcoming of absence, is the first movement of truth. It is the movement which Heidegger termed projection in *Being and Time*. The basis of projection is to be found in *βουλεύεσθαι*. *Βουλε* is resolution; it is the being resolute which constitutes the situation. (cf. Heidegger, 2003, p. 103; GA 19, p. 150) In *Being and Time*, Heidegger tells us that '*resolution is precisely the disclosive projection and determination of the actual factical possibility*' (BT, p., 275; H, p. 298 [emphasis in original]) The evidential in the movement underlying resolution lies the manner in which it shows the situation in its temporal constitution. This is so because the *βουλεύεσθαι* needs a certain span of time. (cf. Heidegger, 2003, p. 104; GA 19, p. 152; cf. also *Nicomachean Ethics*, 1142b3) The latter should not be understood quantitatively. 'As *εὐβουλία φρόνησις* is *genuinely what it is*.' (Heidegger, 2003, p. 102; GA 19, p. 149 [emphasis in original]) In *εὐβουλία*, the concern is to take consideration of time. (cf. Heidegger, 2003, pp. 106–7; GA 19, p. 155) On the other hand, we do not encounter truth in *εὐστοχία* – sureness of aim, instinctual certitude. To act instinctively in a situation requires no consideration because in such a case the absence of the entity is not experienced. (cf. Aristotle, *Nicomachean Ethics*, 1142b2–3) Proper consideration occurs in resoluteness.

In *Being and Time*, Heidegger states that 'in resoluteness the most primordial truth of Dasein has been reached, because it is *authentic*.' (BT, p. 273; H, p. 297) In resoluteness, this situation is disclosed to Dasein. It is disclosed not as objectively present, but as the right time to act, the kairos. In that situation, that which was concealed shows itself as unconcealed. This does not mean that Dasein sees *something* new, but rather that the possibilities of action and knowledge, which are possibilities of Dasein, are opened up in the entities in new and other ways. These possibilities open themselves up as futural because 'what is projected in the primordial existential project of existence revealed itself as anticipatory resoluteness.' (BT, p. 299; H, p. 325) As resolute, Dasein is ready to be 'there' and to act in the situation of failure and of the unconcealement of entities.[18] It acts, however, without certainty, because the situation cannot be calculated or reckoned with in advance. The 'truth of existence' which is revealed in resoluteness is not an eternal truth; it is not even a truth which is as long as there is Dasein. Rather, it is the truth of existence as authentic. It is the truth of Dasein in its coming back to itself. That which is disclosed in resoluteness can be discovered subsequently and can be maintained as true. But the certainty which emerges thereby does not go to the core of resoluteness, because the latter is the response to a situation which remains uncertain, but for which Dasein must remain open in its resoluteness. (cf. BT, p. 284; H, pp. 307–8)

The question arises here as to how this movement of truth relates to propositional truth. By propositional truth we understand the property of a sentence which allows it to disclose an entity as it is. Ernst Tugendhat is certainly correct to argue that a theory of truth which cannot account for propositional truth does not fulfil its most basic task. (cf. Tugendhat, 1970, pp. 331–63) As such, the task here is to see how the movement of truth relates to propositional truth. To do this, we need to investigate the temporality of propositional truth.

Propositional truth remains in the domain of the objectively present and hence of causality.[19] As such, it is to be expected that it can only be understood on the basis of chronological time. Directionality is fundamental to the order of chronological time: it is forwardly directed and irreversible. It is time as measure. In this sense, we can speak of time as standard (*Richtmaß*). (cf. Heidegger, 1998, p. 142; Wm, p. 182) This standard arises with time itself. The clock is its paradigmatic form, but the setting of dates, historical narration and music are all made possible through time as standard. In this sense, time has the quality of normative normality, which is constitutive of it. In German there is a linguistic connection between such directionality (*Richtung*), standard (*Richtmaß*) and the correctness (*Richtigkeit*) of propositional sentences. This connection is not simply a matter of word-play, but rather points to the fact that chronological order of time forms a presupposition of truth as correctness. The relation of correctness is towards that which appears as present. A proposition is true in the sense of correct (*Richtig*) when it is directed in an adequate way to that which is objectively present. (Heidegger, 1998; p. 177; Wm, pp. 228–9) The apparent is the standard for the 'presentative correspondence' (*vor-stellende Angleichung*) of correctness. Here order is understood as actuality, not possibility, because we can only direct ourselves – in the way here understood – towards the actual.[20]

This order is based on the simultaneity of the proposition and that which is shown in it, as two (actual) objectively present things. Only then when there is this simultaneity can there be a correspondence (*Angleichung*). We can take here Heidegger's own example of the verification of a proposition: '"The picture on the wall is hanging crookedly"'. (BT, p. 200; H, p. 217) Although Heidegger does not make it explicit here, it is clear that the person saying this silently expresses a 'now': 'The picture on the wall is hanging crookedly *now*'. For this sentence to be true, the picture must be discoverable now. This simultaneity is chronological. Only in the case that time is dateable can we at all speak of propositional truth. This is so because, when someone states a propositional truth, he is implicitly saying that the thing of which he speaks is present. The thing need not be physically present, of course; the proposition may correspond to an idea such as justice or the square root of 4, but it must be present or represented, and as such available to verify the proposition.[21] This 'now' is, however, not simultaneous with the disclosure of the entity. Disclosure is in the first instance not a matter of the entity, but rather of the being upon which the entity is projected. The disclosure of being cannot itself be dated; it is rather always projected. The being-discovered of the entity is, on the other hand, always dateable. For this reason, the statement 'Dasein is in the truth' does not mean that in an ontical sense Dasein has all truths available to it. It means, rather, that in Dasein as possibility the world is disclosed and as such unconcealment happens. But in that case, the question as to how the movement of truth can appear in truth as propositional is not yet answered, but rather if anything has become more obscure. In order to attempt to shed light on this, we must return to an investigation of the phenomenon of freedom, always bearing in mind Heidegger's thesis that 'the essence of truth understood as the correctness of the proposition ... is freedom.' (Heidegger, 1998, p.142; Wm, p. 183 [translation modified])[22]

II.2 Kairological truth: Free releasement, liberation and truth

Freedom is understood by Heidegger in 'On the Essence of Grounds' as the freedom-to-ground. In the 'Essence of Freedom,' on the other hand, the essence of truth is freedom. Both theses have their roots in *Being and Time* and are not simply two opposing theses which document Heidegger's 'turn from freedom to truth,'[23] rather both are ways to approach the kairology of being. Heidegger's reconception of freedom and truth both in *Being and Time* and in the wake of that work make possible the thought of the kairos and cannot be reduced to chronological concepts. Heidegger's attempt to think 'freedom as ground' and 'truth as freedom' struggles to think ground – hence reason – and truth kairologically.

The roots of Heidegger's thesis that the essence of truth is freedom lie in the phenomomenon of free releasement (*Freigabe*) in *Being and Time*. Quite helpfully, Günter Figal explains free releasement with examples from everyday speech, such as free place, a free street, a free machine. Free means here accessible or open. What Heidegger wishes to say with the concept of free releasement, according to Figal, is that 'the manifestation of entities belongs essentially to Dasein [*die Offenheit des Seienden wesentlich zu Dasein gehört*]' (Figal, 1988, pp. 88–9). Figal is anxious here to avoid the interpretation that Heidegger is appealing to some sort of originary action, and that the relation to entities is produced by something like a transcendental subject. What is crucial for us is the manner in which free releasement relates to practice, that is, whether this releasing into openness occurs originally in poiesis *or* praxis.

Doubtless, Heidegger understands free releasement in *Being and Time* on the basis of handiness. This orientation shifts, however, in the later essay, 'On the Essence of Truth,' and the lecture course of the same name, which Heidegger presented in the winter semester of 1931–2. The basic thought, however, remains the same, namely, that Dasein does not produce the relationship to entities out of its own freedom, but rather Dasein can only experience itself as free in its relation to entities which are freely released, that is opened up to it. It is only because that which is encountered as inner-worldly is freely released for concernful circumspection, that Dasein has the possibility to deal with handy things. The material is not simply 'there' so as to be formed into a work, rather the possibility of the work is formed in the free play between Dasein and its world, in a 'between' which is the place of the letting-be of being. In Heidegger's formulation, 'freedom reveals itself … as the letting be of entities.' (Heidegger, 1998, p. 144; Wm, p. 185 [translation modified])

Heidegger interprets letting be as 'letting oneself engage with entities'. This 'letting oneself engage' means 'to let oneself engage with the open region and its openness.' (Heidegger, 1998, p. 144; Wm, p. 186) This 'letting oneself engage' does not mean that Dasein loses itself in entities; it is rather the case that Dasein steps back in its letting oneself engage before entities. This sounds paradoxical given that the movement of letting oneself engage goes in the opposite direction as the movement of stepping back. Implicit here, however, is the 'logic of failure',[24] as in its everyday engagement with entities Dasein limits its possibilities to entities without reference to the horizon of these entities. It is only when these entities fail that Dasein is compelled to step back and let itself approach the openness itself, hence the horizon of entities.

As in *Being and Time,* Heidegger here encounters the problem of arbitrariness. If truth is freedom, does it not thereby lose its sense? In any case, is truth not 'eternal'? (cf. Heidegger, 1998, p. 143; Wm, p. 184) Stated provisionally, Heidegger's response to this is that truth is not subject to the whims of human beings, because freedom is not a property of the human, but on the contrary the human being is first itself in freedom. This thesis – as already stated – is based on the structure of understanding. Understanding is grounded on temporality, specifically the futuricity of Dasein. Only if the temporality of truth is conceived kairologically can the statement that the essence of truth is freedom be understood in its proper light.

Unconcealment is no longer being understood here – as in *Being and Time* – on the basis of the disclosure of Dasein, but rather as a historical occurrence. Only in the historical moment of vision – when the thinker poses the question: what the entity is – is unconcealment experienced at all. (cf. Heidegger, 1998, p. 145; Wm, p. 187) Being appears for Dasein only in the unconcealment; Da-sein 'occurs' (*geschieht*) only through the unconcealment of being. As such, the disclosure of Dasein has no primacy with respect to the unconcealment of being, because if it is impossible to found historicity in temporality, then Dasein has to be understood as a historical, repeatable occurrence, in which the unconcealment of being occurs as a historical occurring.[25] For Heidegger, in 'On the Essence of Truth', not alone is the unconcealment of being historical, but it is originary to history itself.

> The originary disclosure (*anfängliche Entbergung*) of entities as a whole, the question concerning entities as such and the beginning of Western history are the same and are together simultaneously in a "time" which, itself immeasurable

> first opens up the open region (*das Offene …eröffnet*) for every measure. (Heidegger, 1998, p. 145; Wm, p. 187)

Here a new region opens up, not on the basis of what already was, but rather possibility is opened up as possibility. In this sense, it is not only that a new horizon emerges, but more fundamentally the possibility of the new itself emerges. The new occurrence is not to be explained on the basis of what has already occurred because the measure of explanation is itself opened up by the occurrence. The occurrence is as such 'originary' in an emphatic sense, and as such not explainable in causal terms.

The occurrence of unconcealment does not, however, emerge out of nothing. It happens – as already mentioned – when the thinker poses the question about entities as a whole. Such a question awakens the movement of understanding. Only in understanding is the entity transcended and revealed in projection. The question arises here as to what the truth of projection can mean.[26]Already in *Being and Time*, Heidegger had posed this question: 'Where are the guideposts to direct the projection so that being will be reached at all?' (BT, p. 288; H, p. 312) In effect, this question asks how truth is at all possible in the uncertainty of kairological time.[27] The truth of projection occurs suddenly, happens in the moment of vision, as here time is fulfilled, but at the same time as origins happen, levelling off begins also.[28] It is in this context that the question concerning the truth of projection can be posed. If projection is itself historical, there is no measure outside itself on which its truth can be gauged. The horizon of projection opens up with the projection itself. Such an opening is a breaking off and a breaking out. In this lies the freedom of projection, in the free breach. This free breach happens suddenly as a movement of liberation and self-binding.[29]

The suddenness of the occurrence marks the break with chronology. To say that the occurrence occurs suddenly is to say that it could not be predicted on the basis of what went before, nor could it be explained on this basis subsequently. Plato had described this sudden occurrence in the second stage of the allegory of the cave. (*Republic*, 515c4–e5, following Heidegger's threefold division of the allegory, cf. Heidegger, 2002b, pp. 27–9; GA 34, pp. 35–8) In that context, the occurrence is a liberation and is the first step on the path to truth: the cave-dweller is liberated from unknowledge (*ἀφρόνησις*). The liberation failed, however, because the one who was to be liberated misunderstood the occurrence. He was confused by it and returned to his chains. Heidegger contrasts the suddenness of the occurrence with the descriptions in the third stage of the allegory (*Republic*, 515e5–516e2) of the 'becoming familiar' (*Vertrautwerden*) and the 'capacity to wait' (*warten können*), which characterize the disposition of the freed prisoner. (cf. Heidegger, 2002b, pp. 32, 33; GA 34, pp. 42, 43 [translation modified]) In this sense, there are two moments of the kairos, which are to be understood in terms of the movement of liberating projection.

The freedom of the occurrence presupposes a liberation. This liberation, according to Heidegger, is one which is the ground of history. This 'grounding' of history 'comes to word', not as expression, but rather as 'the ably conserved articulation of the truth of beings as a whole.' (Heidegger, 1998, p. 152; Wm, p. 196) Only in the moment of vision – Heidegger speaks of the 'moment of vision of world' (*Weltaugenblick*) – does this truth happen, and that is so because only as liberation and projection of freedom, hence *between* liberation and being free, does the unconcealment of being happen. It is thus consistent that Heidegger concludes that the question of the essence of truth is the question about the truth of essence. (Heidegger, 1998, p. 153; Wm, p. 198) 'Essence' (*Wesen*), however, means 'being'. If being is to be understood on the basis of freedom, and if unconcleament occurs only as unique repetition, then unconcealment cannot be understood causally, but rather as a free breach. From this the question emerges as to how the essence of an entity is to be interpreted in its truth. (Heidegger, 1998, p. 153; Wm, pp. 197–8) This is a question concerning kairological truth. The old order breaks down in the occurrence of kairological truth. For it to breakdown, the old order must be exposed as in 'errancy' (*Irre*).

'Errancy' does not mean a false doctrine or untrue statements, which could be corrected. Rather, errancy is the forgetting of entities in their totality, which lies at the basis of all activity – it forms for Heidegger the space for normality.[30] It is the normal chronology of the everyday, in which anything new is only comprehensible on the basis of the already 'available intentions and needs'.[31] It is striking here that Heidegger analyses errancy as the possible mode of acting of people in errancy in a manner structurally parallel to his account of curiosity. In Heidegger's formulation, 'The human being's flight from the enigmatic towards what is readily available, onward from one current thing to the next, passing the enigma by – this is erring.' (Heidegger, 1998, p. 150; Wm, p. 194)[32] No more than curiosity (and, as such, fallenness), errancy cannot simply be set aside or corrected.

Heidegger does not place errancy on the level of propositional truth, within a bivalent relation of verification, but rather understands it as the possibility of propositional truth itself, to the extent to which propositional truth intends to reach actuality. As forgetting, errancy is nothing less than the possibility of turning towards the accessible, hence the actual. (cf. Heidegger, 1998, p. 149; Wm, p. 193) What this means is that *within* chronology there is the possibility of going beyond chronology. Such possibility does not depart from errancy, rather it consists in *repeating* errancy. Errancy is experienced explicitly in this repetition. As Heidegger puts it:

> By leading them astray, errancy dominates human beings through and through. But, as leading astray errancy at the same time contributes to a possibility that humans are capable of drawing from their ex-sistence – the possibility that, by experiencing errancy itself and by not mistaking the enigma of Da-sein, they not let themselves be led astray.' (Heidegger, 1998, p. 151; Wm, p. 195)

Heidegger does not refer to repetition here, but only in repetition can the experience of errancy be understood as possibility, i.e. as something which refers beyond itself. The possibility of errancy does not lie at the level of propositions, rather it is the possibility of Dasein to relate to entities in a way only made possible through the unconcealment of being.

Just as we saw in the analysis of curiosity in *Being and Time*, in the occurrence of repetition Dasein is brought before the possibility of the new. Here the possibility of the emergence of a world opens up, which is quite different from the world of 'newest needs and purposes'. (Heidegger, 1998, p. 149; Wm, p. 193 [translation modified]) In the opening of possibilities in projection emerges at the same time a measure of truth. This would, however, appear to be a non-sequitur, namely, that of calling truth that which is at most the conditions of possibility of truth. As is well known, Ernst Tugendhat accuses Heidegger here of a logical mistake. (cf. Tugendhat 1970, p. 297) As Daniel Dahlstrom has shown, this charge can be denied if it can be shown that the propositional truth is co-constituted by the temporality of its emergence.[33] In this way, a second critique of Tugendhat's can also be rebutted, which can be reformulated in terms of the present book: in kairological truth, the specific concept of truth is lost, that is, the disclosure of an entity *as it is in itself*. In doing this, an answer will be given to the question already formulated as to whether there is a connectedness between propositional truth and the truth of kairos. (cf. Tugendhat, 1970, pp. 332–7)

In the kairos, the entity does not show itself in its actuality, but rather it is disclosed in its possibility, that is, it is manifest in relation to being and Da-sein. What is thus brought to appearance is the *order* in which the entity can be meaningful. Such order is historical in the sense that the possibility belongs to it of transitioning into the new. Without this possibility of transformation, there would be no truth. Truth, understood as correctness, presupposes the possibility of transformation, in which the transformed order of entities makes a claim on Dasein to correspond to it. Without such a transformation, there would be no ground underpinning why Dasein should direct itself correctly towards entities: it would encounter entities only in their accessible serviceability; the gulf of absence, which we have already seen to be a characteristic of truth, would be lacking.[34] Only due to the instability of any particular order is the question of truth posed. That we can nonetheless conceive of propositional truth as a-temporal does not speak against the basis of such truth in historicity.

The question remains, however, as to the extent to which propositional truth is co-constituted through kairological truth. For this to be the case, kairological truth must not simply be the condition of possibility of truth; it must also condition the structure of this truth. In other words, propositional truth must itself show certain kairological characteristics. We can take a simple example, such as 'the window is open'. This sentence refers to the actual – the window – but does not confine itself to the realm of the actual: it refers also to the possibilities of the actual – the window is open is true only if it can be false: if the window can be closed. The actualization of one possibility – in this case the opening of the window – occurs in normal, chronological time. These possibilities are such in the entity for Dasein only because of the latter's possibilities which allow it to relate to such entities. There are, of course, windows which cannot be opened. As such the sentence 'the window is open' cannot be true of all windows. As I sit at my desk in my study, I can open the window in front of me; sitting in the library in front of sealed windows, this is not possible. The order of my world has shifted somewhat. The sentence 'the window is open' in the latter case is not only false, it has not the possibility of being true. On the other hand, the sentence 'the window is closed' is also not true, as the window has not even the possibility to be open, and is, strictly speaking, as little closed as, or closed only in the sense that the walls of the library remain intact. In such a case, the sentence, 'the window is closed' is meaningful, but is neither true nor false. This is, however, a strange conclusion as it goes against the thesis, which goes back at least to Aristotle, that a propositional sentence must be true or false (the principle of bivalence or the excluded middle).[35]

This consequence is, however, in line with the considerations concerning contingency at the beginning of this chapter. We saw there that the exception to the principle of bivalence which Aristotle himself saw – concerning the future – concerned all three temporal dimensions, when they were considered kairologically. *Normally*, the principle of bivalence is applicable: the window is either open or not in actuality. Here, though, with respect to the possibilities of Dasein in its relations to the window, two things are assumed: that these possibilities are clearly presupposed and, at the same time, forgotten. It is a totally different situation, however, when these possibilities are encountered by someone for whom they are strange and unfamiliar. Such a person is not in a position to make any true propositional sentences that he can verify to be true. That does not point only to a lack of knowledge. The experience of strangers shows that a familiarity with certain entities which are the objects of propositions is essential to the successful employment of those propositions, and that to this familiarity there belongs possibilities of relating to those entities. A proposition is only true, if it discovers entities in such a way as they present themselves *in these possibilities*.

In the moment of vision of liberating projection and the encounter with a strange world, Dasein has no object of consideration in the manner in which it is familiar – in an unfamiliar world things seem unreal, almost as if what we see were only a façade, without substance. In such a case, Dasein is concerned only with the order of entities. The freedom of the stranger, who can deal with entities in ways other than those of the native, is the freedom of the letting herself approach the letting be of being. In this free action, the stranger encounters indeed the boundaries of entities, but these boundaries are often different from what 'one' thinks. The propositional truths which can be uttered in such a case discover the entity as it itself is, but discloses it in a manner unfamiliar to those who are familiar with it. If, on the other hand, the proposition is not true, then the relation to the entity can also not be fulfilled. What this shows is that actuality is not an arbitrary moment of possibility, but a necessary one. The possibility of dealing with an entity cannot be separated from the reality of this entity: the possibility to open a window must coincide with the experience of a reality – the window as actually capable of being opened.

Kairology and chronology cross over each other in truth. The occurrence of truth as a free, kairological occurrence liberates Dasein from the concernful dealings with entities, so that it can respond to the transformation of order and through that with the emerging of time. Just this freedom is the essence of chronological truth. This truth is not that of the causal world; it presupposes freedom and transformation: in order that Dasein can relate to an actual entity, time as the making possible of the order of entities, must have already emerged.

III The emerging of time

Time emerges in the free contingent moment of vision of the unconcealment of being. This emerging occurs between liberation and being free in a free releasement, which fulfils itself in action, but which can never be the goal of action. The emerging of time is not something which can be made actual; it is a possibility of transformation, which can never be contemporaneous with a statement or an action. As such, the emerging of time can never be directly experienced because that which can be *directly* experienced is that towards which we can direct ourselves. The occurrence of this emerging is something which can be felt in its intensity.

'Intensity' is not a term typical in Heidegger. When Heidegger speaks of the mood of failure – hence, presumably, of an intense experience – what he stresses is the indifference of its environment for Dasein. But what is indifferent are the inner-temporal entities. Temporality itself is not indifferent, rather Dasein experiences itself as temporal.[36] This experience of temporality is both painful and pleasurable.[37] What we find here is an affective intensity, which is not simply a subjective reaction to time, but is constitutive of time itself. This intensity cannot last: the human being cannot bear such intensity – neither corporeally, nor spiritually – for longer than a limited duration. The significance of this time, however, does not lie in its duration. Within many mystical traditions this intensity is interpreted as an entrance to eternity. Michel Theunissen interprets this intensity as lingering (*Verweilen*) and attempts to gain a philosophically legitimate concept of eternity on the basis of such experience. (Theunissen, 1991, pp. 285–98) Against Theunissen's account, which is rooted in the claim that human beings strive for freedom *from* time, I wish to argue that freedom is achievable *through* time, namely in the kairos.

It is significant that Theunissen takes the visual arts as a paradigm to support his account of lingering. He understands, of course, that this choice is one which is oriented toward supporting his argument: 'a negative relation to time belongs to the visual arts just as essentially as a positive one does to music.' (Theunissen, 1991, p. 287). In music and in dance,[38] we find examples of the structurally constitutive function of time in making them possible. Furthermore, these are examples not only of a positive, but also of an intensive temporal relation. The experience of rhythm is one in which we (to use Theunissen's terms) 'go along with time' (*mit der Zeit mitgehen*) (Theunissen, 1991, p. 285), and we do so in an intensive manner. Time as we encounter it in rhythm is no longer dateable and its significance cannot be understood in terms of the in-order-to structure of handiness.[39]

Rhythmic time has, however, its own order and power. Precisely through rhythm we are freed from chronology, and this can lead us even into 'chaos'. In giving ourselves over to the rhythm, we lose power over ourselves and are placed under the force of the rhythm. Through the intensity of rhythm, normality falls away and the ecstases of time close together. The energy of this order appears chaotic, as the constant energy of the bringing-into-order comes to appearance in its intensity. We can sense the destructive power of this energy in the movement of rhythm. Above all, Dasein experiences the power of time as not being in its control. Dasein can indeed make up a rhythm, but only because it already lives 'in' rhythm. It can use rhythm for a particular purpose, but only because it has received it from elsewhere.[40] In rhythm, the liberating is also a force, a way to chaos and at the same time a strict order. Therein lies the ambiguity of kairos: the extraordinary time, which is itself an order of time. This ambiguity is at the basis of that between happiness and misfortune, which we have seen to be characteristic of the kairos. In the kairos, Dasein hovers between order and chaos; it acts within a free play of possibilities, in which there is no standard, within which, however, its action takes place under the force of the rhythm of the moment of vision.

The examples, or better analogies, of music and dance give two clues as to the characteristic time of such action under the force of rhythm. Firstly, in music and dance there is no time for reflection; if during a dance or in performing a piece of music we reflect upon what we are doing, then we lose the rhythm and fall from the intensity of time. This lack of reflection is characteristic of time itself: when time has such intensity, then there is no time for reflection, but rather only time for response and for going along with it. Secondly, the power of time is experienced here. In inauthentic dealings with time, the power of time is also experienced, but in a different way. The intensity of time is the intensity of its power. Such power is not to be understood – as Theunissen would have us believe – as a dominating power, but rather as a power which runs through Dasein. It comes over it; it is not in its possession, but it makes it possible for it to do that which otherwise would be impossible for it. Without rhythmic time, which encompasses her through music, the dancer cannot dance.

In music and dance, it is not the passing, but the *emerging*, of time which appears. The future does not simply come to Dasein, but springs up before its eyes and draws it in. This enticing power of time makes it possible for Dasein not simply to go on living, but rather, in a new origin of time itself, to 'begin' anew.

The kairos takes place as pure contingency, which is a matter not only of the moment of vision, but also of the having-been. That which has been is experienced as possibility and is transformed in relation to the moment of vision. This occurs in a struggle with the past. This struggle is directed at the chronological order itself. It is a struggle against the necessity of the past, which Heidegger had characterized as 'thrownness'. The struggle against this necessity makes possible the emerging of time. In a word: time emerges out of a struggle between chronos and kairos. That both of these are reciprocally dependent on one another is demonstrated in this struggle.

We can measure the rhythm of a piece of music or poem; without chronological time, it would be impossible. Nonetheless, it transcends chronological time. Rhythm does not simply occasionally transcend chronology, it is inconceivable without both chronological and kairological time. This is so because rhythm makes it possible for Dasein to go along with the power of the emerging of time in the moment of vision. When it is said that in revolutionary situations time goes at a faster rhythm (Koselleck, 1985, p. 77), this is not to be understood merely metaphorically. In everyday life, we experience the rhythm of time differently, say, during a workday and while on holidays; in a similar way, the rhythm of time is experienced differently in times of stability from that of times of unrest. In order to act in a manner appropriate to each situation, we need to go *with* the rhythm of time.[41]

The kairological moment is not an abstraction from life, is not a lingering with the eternal, but rather a giving of the self into the possibilities of living, which are for the most part suppressed. This giving of the self into life is characterized by the intensity of time. We can understand this intensity with the help of Jasper's concept of 'enthusiastic attitude' (*enthusiastische Einstellung*). (cf. Jaspers, 1989, pp. 117–36) According to Jaspers, love is an example of an enthusiastic attitude. Love, he understands as a struggle, in which the participants are without power. What this shows is that struggle does not always involve empowerment, but can also be an 'expression of a process of the *intensification of understanding* in love.' (Jaspers, 1989, p. 126 [my emphasis]) The struggle so understood has no goal outside itself; its goal is itself: its tension is at the same time its fulfilment.[42] It forms an intensification of understanding. One is no longer satisfied to understand the other (the beloved) through 'normal' (and abstract, general) interpretations, but rather is ready to wager one's own and the other's life, and thereby to understand that life in its individuality. In such wagering, we experience a 'unique freedom.'[43]

In this intensification, freedom takes effect, precisely when the human being no longer has power over himself, when he no longer asserts himself, but rather gives of himself. In this self-abandonment, we find freedom *from* the general and *for* the individual. (Jaspers, 1989, p. 124) What is at issue here is not any question of individualism, but rather concerns the problem of individuality. In the kairos, we are concerned with a moment in its uniqueness. It is not accidental that the moment of freedom is also unique, because freedom is not reducible to the past, but is rather a breach with the past, a breach with the attempt to understand individual events in a comprehensive history as expressions of a general, teleological chronology. There are, of course, fundamental problems in bringing the unique to linguistic expression. When we attempt to do so, we tend to lose it because the very words we use express the general, not the individual. Heidegger attempts, through a phenomenological analysis of mood, to find a way of discussing the unique. Mood is precisely speechless.[44] In mood, Dasein is discovered as possibility, hence as unique, as a possible place for the new. Only in mood can this uniqueness be disclosed.[45] In mood, Dasein comes back to itself as possibility.

Kairos is the temporality of essential action. Dasein is speechless in the kairos, not because it has nothing to say, but because it wishes to speak but cannot come to language. Dasein must respond to the unique, but it can do so only through a leap, which forms a breach with causality. As resolved, Dasein makes this contingent leap because resolution discloses the indeterminacy of the situation. The leap is a dance over the abyss in which time is experienced in its intensity. The rhythm of time is experienced as ecstasis, that is, as abysmal – groundless – movement. In this intensity, there is an order, namely rhythm. This order is a transformation, an order without ground, which can transform itself in any number of ways. In it, the free releasment of being happens: a new order emerges.

Such emergence is not, however, directly experienced. Just as birth is not experienced as such, so it is also with the event of emergence. Once the kairos is recognized, it is already past. Dasein acts in it, but that action becomes itself an instance in time, which can be dated and which can be narrated as a fact. Such narration takes action as already past, as a part of a history. Every outbreak of the new is at once actual in its effects and can be chronologically ordered. In chronology, freedom is subordinated to causality. When a historian – or indeed the participant in recollecting an event – wants to give a history of a revolution, war or some major change, or when a lover tells of the time in which she fell in love, both will attempt to integrate that event into the past through linking it to the causes which brought it about.

Action in the kairological situation is not only an *occurrence* of truth. It lies in the essence of truth that it be preserved.[46] What is at issue here is the *grounding* of a new order. In this experience of grounding or establishing, the question of the possibility of grounds is posed. Grounding is a poietic act.

Time as emerging must be preserved. Without such preservation, Dasein would have no identity and no history. Preservation occurs in grounding and this raises anew the question of the relation of poiesis and praxis. In now turning to Heidegger's discussion of art, we do so because the inner relation of transformation and preservation, of possibility and actuality – the ontic and the ontological – temporality and historicity, still remains obscure.

4

The Time of the Work I: Art

'The actualizing (*Wirkung*) of the work (*Werk*) does not consist in an effecting (*Wirken*). It lies in a transformation (*Wandel*) of the unconcealment of entities which occurs out of the work, a transformation, that is to say, of being.' (Heidegger, 2002, p. 45; Hw, p. 58 [translation modified]) These words from 'The Origin of the Work of Art' indicate the crucial issue, which Heidegger is addressing with respect to the work. The work is actual (*wirklich*) but cannot be understood in terms of effects (*Wirken*). Furthermore, in and out of the work occurs a transformation in truth. One can justifiably conclude from this that there are kairological characteristics in the work. Kairos, as we have seen, is nonetheless in possibility. But for Heidegger, the work comes to be seen as the 'place' of the interplay of actuality and possibility and this complicates the relation between, on the one hand, kairos and chronos and, on the other hand, praxis and poiesis. This chapter and the next will investigate the place of the work as Heidegger develops it in the 1930s in relation first to art and then to politics.

According to Otto Pöggeler, Heidegger's question concerning art must be understood as a question concerning its possibility in the technological age. (Pöggeler, 1969, p. 46) But this is to read Heidegger too retrospectively. Heidegger had already written the first draft of his Artwork Essay in 1931 (cf. Heidegger, 1989a, p. 5), at a time in which technology was not yet a theme in this thought.[1] If we read Heidegger along the trajectory of this thought, at issue is, rather, the question of the possibility of action as it relates to actuality. Art is a possibility of human action, which apparently is actualized in the artwork. However, as we have shown, for Heidegger in *Being and Time* the possibility of Dasein is precisely not that which can be understood in terms of its actualization. Can art be understood in relation to this account of possibility in Heidegger's magnum opus? Or more fundamentally still: can the account of possibility in that work account for actuality as the condition and the product of human action?

Philosophy understood as reflection on the happening of ontological difference, can never ground this happening. Philosophical thinking reaches towards that which transforms itself, a transformation which, however, occurs without a *subject* of change. Philosophy suffers from its passivity in relation to the happening of the ontological difference, which it thinks but cannot ground. The fundamental question of philosophy then concerns freedom. (cf. Heidegger, 2002a, p. 203; GA 31, p. 300) Freedom must be understood here not as the exercise of power, but as that which is an issue at the boundaries of philosophy's possibilities, precisely in the failure of the attempt to found the ontological difference. Philosophy neither founds nor grounds anything.[2] As such, we misunderstand the essence of philosophy when we understand it as poiesis. The question which arises then is what the truth of philosophy is, that is, what its mode of disclosure is.

With this question, the problem of the relation of praxis and poiesis has to be posed again from a new perspective. In the wake of *Being and Time* Heidegger gives increasing weight to poiesis – the 'work' becomes a central theme. With this, the relation of kairos and chronos changes. In the wake of *Being and Time*, the question of the possibility of historical action, in particular with respect to its temporal structure, comes to be an increasingly important theme. In this context, philosophy itself is called into question: if it is the case that philosophy cannot ground history, but always through its destructuring reaches towards the unique happenings of the ontological difference, then philosophy is both the knowledge of historicity and itself a historically conditioned form of knowledge. Any such, historically conditioned knowledge is a form of action: it takes place in the interplay of the movement of the happening of history. If this is the case, then philosophy always needs to secure its own mode of relating with and against other possibilities of action. Heidegger's political engagement – as controversial as that is – must be understood as the attempt to determine the possibilities of action of philosophy and politics in their reciprocal relations and also to draw out the implications of a kairological understanding of action. For Heidegger, both politics and art/poetry concern the relation of philosophy to the unique and unrepeatable character of the occurring of the transformation of entities.

I Historical action

I.1 Praxis and poiesis after *Being and Time*

In the terms laid out in *Being and Time*, historicity comes first to appearance for Dasein when it encounters a piece of equipment from a past world. (cf. BT, pp. 348–9; H, p. 380) A piece of equipment which is displayed in a museum is past, but still present. The piece of equipment is objectively present, but it is no longer at hand (*zuhanden*) as a piece of equipment. No disturbance of the piece of equipment is necessary for it to be perceived as objectively present. But something has been disturbed, as the piece of equipment does not as such belong in the museum. But this piece of equipment, even if it still functioned, would no longer fit in with today's work-world.[3] It is, nonetheless, actually a piece of equipment; it has been used. We know that it would not be used in the world of those who visit the museum. But a world comes to appearance in the piece of equipment, which is no longer, but was once. This is what Heidegger means when he poses and answers the question: 'What is "past"? Nothing other than the *world* within which they [the tools] were encountered as things at hand belonging to a context of useful things and used by heedful Dasein existing-in-the-world.' (BT, p. 348; H, p. 380) This world is primarily a work-world. If, however, historicity consists in Dasein's capacity to repeat the modes of relations which have-been, and if Dasein is itself called to account through this repetition, and if further the world thus repeated is the work-world, then it seems to follow that Dasein's experience of its historicity occurs through making, that which is made and as that which leaves behind the trace of its making in its works. But how do we go from there to the conclusion that the human being as Dasein – unlike any other innerworldly thing – can never be past, but always 'having been there'? The world of the tool – the work-world in which Dasein recognizes itself as historical – is itself past. It follows from Heidegger's thesis that historicity is only a working out of temporality, that Dasein can never be simply past, because the being-a-whole of Dasein is preserved in its temporality. If the results of Chapter 1 are accepted, namely that this thesis does not hold, then the question arises as to how Dasein can experience itself in its connectedness with the past world, given that the connectedness of its life cannot lie prior to its birth. In other words, how does Dasein identify itself with a world which is since past? How in its experience does an old piece of equipment, an old tool, with which it has no familiarity at all, differ from a thing which is not made by human beings? When we encounter, for example, a primitive stone-age tool, often the question can arise as to whether it is in fact a tool or rather simply a stone, which accidentally has a shape that makes it appear capable of being used as a tool. It would be to pervert Heidegger's meaning if we were to claim that this can be decided on the basis of scientific research. Such things are indeed displayed in museums, but it cannot be simply on the basis of the authority of the museum curator that Dasein can let the world of the tool be opened for it.[4] The more elaborate the tool, the less the chance that a mistake is made regarding its nature as a tool. But even then Dasein does not recognize itself but, at most, the peculiarity of a highly developed animal belonging to the human species.[5] In the tool as such, Dasein does not recognize that which puts a claim on it. In the tool in the museum, a human world indeed opens up – although as just suggested even this may not occur – but this world is as yet not a world of Dasein. That it is a world of Dasein – a past historical world with which Dasein can identify as the having been world of its people, of its community – cannot appear to it out of the tool itself.

Nonetheless, Dasein is bound to that past world, if only indirectly, through its birth. Through its birth Dasein is related to a past, which it had not itself experienced, but with which it is bound in various ways. The individual Dasein can find itself in a historical situation, that is, exist historically, when a time appears as outside its appropriation, not in the sense that it has power over it, but rather that it is strange to it. (cf. Sommer, 1990, p. 146) Heidegger to a certain extent captures this strangeness with the term 'thrownness'. The past makes a claim on Dasein; but this time remains strange to it. At first sight this appeal to the strangeness of such a time does not seem to help us any further, as what is of concern is precisely the extent to which Dasein 'indentifies' with this time. But, as we will see, it is precisely the strangeness in this time which makes a claim on Dasein. A piece of stone which appears as resembling a tool is not strange to Dasein, but at most appears peculiar.[6] What is strange to Dasein makes a claim on it because it is a possibility of Dasein to be other than it is, and in its strangeness this possibility calls on Dasein to identify with it.

In *Being and Time*, Heidegger does not speak directly of this strange time. On the contrary, that which we here attempt to explain with the concept of strangeness, Heidegger discusses with the concept of generation. But there is no necessary contradiction here, at least if we understand the relation of temporality and historicity otherwise than in *Being and Time*. The concept of strangeness does not appear in Heidegger's analysis because historicity is being understood in terms of the transcendental structure of Dasein's temporality. When, however, temporality is thought historically and not as the ground of history, then the past can be explained on the basis of the historical having-been of Dasein.

The strangeness of the historical having-been is in any case quite reconcilable with the generative moment. That Dasein is thrown into the world does not mean that it is thrown into a *historical* world. As we have already seen in Chapter 1, not every human world is historical. In a historical situation, the world in which Dasein is born offers possibilities, which precede it and which are offered precisely as *having-been possibilities*. These are possibilities of a people (*Volk*) – destiny (*Geschick*) – which, however, are such that they have to be repeated (*wieder-holt*). In a historical world, identification with these possibilities is not simply a process of appropriation in which the possibilities cease to be strange, but at the same time an alienation because, in this appropriation as repetition, a transformation occurs. As distinct from, say, a totemic world, though, the strangeness of the appropriation of its past possibilities is not cancelled out, but rather leads to an alienation of one's own world through a transformation, which is presaged already in birth.

There is no need to repeat the discussion of birth and destiny in Chapter 1. What is important here is to see that, from this perspective, the historical world is only secondarily a work-world. Primarily, it is a world of being-with, of transmission between generations, a transmission in tradition which can only occur in the public world. Only in the world of praxis can there be 'heroes' who can be chosen or rejected. (cf. BT, p. 352; H, p. 385) The work-world is not a world of 'choice' or of 'repetition', but rather one of the application of models. In the work-world, Dasein relates to a past, but in this case a past which lies behind it and which can appear to it only those works which are formed on the basis of past patterns. In the world of praxis, Dasein is brought before the singular possibilities of its situation. Heidegger indicates this singularity in the concept of fate (*Schicksal*). (cf. BT, pp. 351–2; H, p. 384) The finitude of the historicity of Dasein consists in the singularity of its historical situation.

However, the manner in which Heidegger thinks the relation of historical praxis and poiesis in *Being and Time* remains in the end unsatisfactory, because he did not pursue the theme of generation. Rather, he continuously refers back to the (supposedly non-historical) temporality of Dasein to explain the foundational relation between the historicity of Dasein and world history. For example, he says, that 'the historicity of Dasein is essentially the historicity of the world, which on the basis of its ecstatic and horizontal temporality, belongs to the temporalizing of that temporality.' (BT, p. 355; H, p. 388) Historical Dasein is understood as a happening of world, but this happening of world happens in Dasein itself, i.e. as founded in the temporality of Dasein as being-towards-death. Birth and the phenomenon of generation, the happening of the temporality of Dasein, which is not in each case mine, is not pursued further. It is for this reason that when Heidegger turns to world-history (cf. BT, pp. 354–8; H, pp. 387–92), he can once again view world as work-world.

The singularity of fate characterizes the belonging of Dasein to a specific people (*Volk*).[7] This belonging means that Dasein always occurs with other

Dasein, and Heidegger calls the happening of this occurrence destiny (*Geschick*). Due to the singularity of this fate, Dasein cannot simply take examples, or even less laws, from the past in order to apply them in the present. Its singularity and the movement of its fate belong together. If this is the case, how do we even sense this movement; how do we sense the trace of it in a past world in the very occurrence of repetition? Furthermore, if history as happening is enigmatic movement; philosophy, in understanding that movement, can do so from no other place than within that movement itself. This is all the more so because philosophy is in a constant return to the groundless ground of movement. In this sense, philosophy is not something other than the movement of history, but rather that very movement in its most accentuated form, that is in its return to its own original force. But movement as such remains incomprehensible, and indeed the movement of return would become impossible in history's moving force. Philosophy requires something else which brings movement – movement in its abysmal historicity – to a stand. To understand what is at stake here, we need to explore two key terms in Heidegger's discussion of art, transformation and preservation.

I.2 Transformation and preservation

Heidegger associates transformation with those modes of human behaviour which prepare for the kairos: in the late 1920s 'philosophy' and 'action' (cf. Heidegger, 1984, pp. 106, 221; GA 26, pp. 132, 285; Heidegger, 1995, pp. 57, 68; GA 29/30, pp. 86–7, 102–3), in the 1930s still philosophy and in addition 'politics', but above all 'art' and 'poetry'. (cf. for example, Heidegger, 2002, p. 40; Hw, pp. 52–3) In each of these modes of behaviour historicity is at issue; transformation is understood as a historical occurring. This the first hint that the project of fundamental ontology will fail, if it is conceived as an attempt to reach a non-historical basic experience underlying the history of western philosophy through a destructuring repetition of this history. A first indication of this failure is to be found in the controversial passage concerning the 'Idea and Function of a Fundamental Ontology' in his 1928 lecture course *Metaphysical Founding Grounds of Logic*. (Cf. Heidegger, 1984, pp. 154–5; GA 26, pp. 196–202) In the course of this discussion, Heidegger states the following: 'Fundamental Ontology is always only a repetition of this ancient, early [manifestation]. But what is ancient gets transmitted to us by repetition, only if we grant it the possibility of transformation.' (Heidegger, 1984, p. 155; GA 26, p. 197) Through repetition, that which is transmitted is transformed. Repetition fails when this transformation does not occur. For Heidegger, 'tradition' does not allow for such transformation. (Heidegger, 1984, p. 155; GA 26, p. 197) For this reason, the repetition of the 'ancient and early' is a destructuring of the tradition. 'Tradition', Heidegger tells us in *Being and Time*, uproots the historicity of Dasein through 'ahistorical categories such as types, directions and standpoints'. (BT, p. 19; H, p. 21)[8] The goal of *Being and Time*, however, is not to transform the ancient and early, but to ground them, to 'fix their boundaries'. (BT, p. 20; H, p. 22) Repetition allows the hidden ground of the repeated, the experiences which lie at this basis, to come to appearance. Through destructuring and repetition a non-historical ground of history is thus revealed. According to Heidegger: the temporality of Dasein and the temporality of being.

There is, however, a contradiction at the heart of this project: repetition which transforms brings to appearance that which cannot be transformed, a fundamental experience. The latter appears to be a 'remainder' in the repetition, which remains invariant in every transformation. But is this result anything other than the 'passing on (*Weitergabe*)', which Heidegger characterized as the failure of repetition? The originary experience is not spoken of in the 1928 course as the historical world in its strangeness, but rather the 'possibility of temporalizing new origins'. (Heidegger, 1984, p. 155; GA 26, p. 198 [translation modified]) These possibilities form the historicity of Dasein.

Historicity does not simply refer to the characteristics of the historical situation of Dasein when it relates to the 'ancient and early'. The transformation in question concerns the whole of Dasein as that entity which finds itself in the ontological difference:

> Since being is there only insofar as entities are already there [*im Da*], fundamental ontology has in it the latent tendency toward a primordial, metaphysical transformation which becomes possible only when being is understood in its whole problematic. The intrinsic necessity for ontology to turn back to its point of origin can be clarified by reference to the primal phenomenon of human existence: the entity "human" understands being; understanding-of-being effects a distinction between being and entities. (Heidegger, 1984, p. 156; GA 26, p. 199)

The ontological difference and the understanding of being happen at the same time. However, this 'happening at the same time' is not to be understood chronologically, as a determinate point in time, but rather as a transformation in which the everyday order falls away and Dasein finds itself without orientation in the midst of entities. It is a moment of vision (*Augenblick*) and recalls the analysis of mood in *Being and Time*, but one which no longer concerns a return to the a priori perfect, but a transformation affecting both Dasein as well as the having-been possibilities: 'transforming the humanity of us human beings into the Da-sein in ourselves'. (Heidegger, 1995, p. 350; GA 29/30, p. 508)

Dasein is the between of the ontological difference, and when a transformation happens in this difference – a change in the being of entities – Dasein is transformed. Being is precisely not to be understood as eternal then, but as a change and movement into presence as present and at the same time beyond presence.

In the word '*Verwandlung*' (transformation) we can hear '*wandeln*' (change) but also '*wenden*' (to turn) (cf. *Duden Herkunftswörterbuch*, pp. 742, 761) One can say: 'the weather has turned', in other words, 'it' has become other than it was. But if we asked *what* has turned, *what* has changed, it is the weather. Weather, however, is not something which can be found. What we experience is rain and then sunshine, a rainy day turns to a sunny one; that is weather. Weather brings itself to bear in rain and sunshine. We say *it* is raining, *it* is warm, *it* snows – in each case something *happens*. It rains – perhaps suddenly or gradually – but it happens as different to sunshine or snowfall. This difference is weather. If it only rained, there would be no weather. Only because there is a difference between rain and sunshine, snowfall and sleet, *is* there weather. This difference takes place as transformation. Transformation here is not a transition between two fixed states of affairs; without this transformation there would be no rain, snow or sunshine as weather. Transformation is a happening in which that which was becomes transformed, yet remains, but not as identical or as a thing. What remains is the weather – weather is experienced as that difference which maintains itself and lets otherness happen.

This example teaches us that transformation is not the movement of a subject or a substance but rather movement in itself; being does not move, it is rather the movement of entities as a whole. 'Weather' is only through an 'it', which is not a something, but rather that which makes transformation possible. The question as to this 'it' is, for Heidegger, the question of philosophy. The concepts of philosophy have for this reason no determinate content because they do not refer to the objectively present, but only indicate 'that anyone who seeks to understand is called upon by this conceptual context to undertake a transformation of themselves into their Dasein.' (Heidegger, 1995, p. 297; GA 29/30, p. 430) To perform this transformation is not an act of Dasein from itself, but is a response to the transformation of being. What is crucial here is to understand philosophy as action. (cf. Heidegger, 1995, pp. 57, 68; GA 29/30, pp. 86–7, 102–3) Philosophy is action in that it brings Dasein, in the finitude of its existence, to clarity concerning itself.[9] What is meant here is not that Dasein encounters its end, but rather what is at issue is the preparedness for the happening of being which occurs in each case uniquely.[10]

In the *Fundamental Concepts of Metaphysics*, Heidegger speaks of a preparation for transformation, which in contrast to birth is not rooted in the continuity of a people or of generations, but involves a transformation of humanity into Da-sein. Philosophy cannot effect this, but can only prepare for it. (cf. Heidegger, 1995, p. 352; GA 29/30, p. 512) In *Being and Time*, Dasein is also not identical with human being, but here it becomes explicit that Dasein presupposes a transformation, which unlike birth does not confirm the human being, but rather transcends it. At the heart of the issue here is the distinction between the strange and the new to which I have already alluded. Transformation is here understood as having nothing in common with the continuity of generations. It is in relation to the past, but this relation is one of over-turning, not of renewal. Novelty is normally understood as that which was not before. But novelty can also refer to a specific relation to the past, namely that of renewal. With the new generation the human race is, we say, renewed. The capacity to begin of the child is a possibility of renewal. However, transformation can be understood as *metamorphosis*: what was and what is become other. In the transformed present, we can no longer recognize the past. But we can still sense it in its absence, as an alien presence in the present. The past haunts the present as an alien body. While the 'rupture' of birth establishes a continuity through the renewal of humanity in its dis-continuity, understood as metamorphosis transformation confronts the present with the alienness of the past, which resists all attempts at integration.

In the wake of *Being and Time*, birth can no longer play a central role. This is so because the question Heidegger increasingly poses concerns not so much Dasein in its modes of disclosedness, but rather the happening of the ontological difference. Transformation is the movement of this difference and the question now becomes one of the mode of action of Dasein in which this transformation happens. This does not mean that the analysis of birth loses all significance. Through birth, Dasein is first historical, and as historical, Dasein cannot cease to engage with the strange past world out of which its world emerged and to varying degrees to identify with it. But the ground of that historicity, indeed the very meaning of ground here cannot be understood simply by reference to the natality of this being.

Heidegger does, nevertheless, appeal to a birth in discussing the transformation of history: not of a human being, but of a 'demi-god.' Heidegger's understanding of this latter concept is mediated through Hölderlin. The account of the 'demi-god' concerns the relation of humanity to divinity. This latter relation has no fundamental importance in the project of *Being and Time*. In 'On the Essence of Grounds', Heidegger claims that the question of God can only be adequately stated on the basis of a firm foundation in the ontology of Dasein. (Heidegger, 1998, p. 371n. 62; Wm, p. 157n. 56) When first in his Rectoral Address and then more fully in the first Hölderlin lectures of 1934, Heidegger speaks of god and gods, his interest does not lie in theology in a Christian sense.[11] Heidegger's concern is not with the questionable presence of the Christian God, but with the absence of the gods, their abandonment of humankind. This absence – the death of God, to reference Nietzsche – cannot be brought to light with the help of an ontology of Dasein, because it is a historical happening which befalls Dasein in its essence. Through this happening of abandonment, Dasein experiences itself in the ontological difference as abandoned by the gods. (Heidegger, 1985a; p. 474; R, p. 13) This abandonment characterizes Dasein's being delivered over to the transformation of being. The flight of the gods is, namely, a happening of being through which entities as a whole are transfigured. (cf. Heidegger, 1989, pp. 97–100) Through this happening, Dasein is thrown into the interplay of presence and absence. The gods have flown, but in the human ability to transcend itself in questioning there is a receptivity for the trace of the flown gods, a trace which comes from the demi-gods.[12] Dionysus, the 'most excellent demi-god', 'presencing this demi-god essences away and absencing he essences near [*anwesend west dieser Halbgott ab, und abwesend west er an*]' (cf. Heidegger, 1989, p. 189). He is the difference between being and non-being, between humans and gods. Because he is such a 'between being', it lies in the essence of the demi-god 'to always be an other' (Heidegger, 1989, p. 280).

There is no underlying order with respect to this demi-god, he is exclusively a between being which is always otherwise. Only in his absence is he present. He is a mask, behind which there is no face. In Dionysus, we do not encounter the familiar face of a human child, but rather only a mask. Entities receive a new figure and this happening of being Heidegger calls 'fate' (*Schicksal*) in his Hölderlin lectures.[13] Fate is a power which dominates Dasein; a power of transformation, metamorphosis. The problem of historical action is first to recognize, and respond to, the power of this metamorphosis.[14]

Transformation as such, however, is unrecognizable; only in its being brought to a stand does transformation become apparent. Only through such a coming to a stand in transformation is there the ontological difference, because this difference is only in and through that which remains. Being is the movement of entities; the between of being and entities is the there of a transformation and the resting place in which it can be traced. In transformation, the past appears in its radical alienness. The bringing to a stand of this transformation is that which allows entities to be transformed, because movement can only be accomplished in repose.[15] This repose is accomplished in the bringing to stand in a work. For this reason, philosophy alone cannot understand transformation without the aid of that which brings it to a stand, namely the work: philosophy cannot think transformation immediately, it can only understand it through reflection on the work, especially the artwork.[16]

There can be no transformation without preservation (*Bewahrung*). This is apparent in relation to the work: the making by the artist, and the 'simultaneous' preservation of the work by the one who lingers before the work.[17] If the work is the coming to a stand of transformation, movement shows itself most fully in preservation. Preservation does not mean a taking possession nor does it mean the reception of the artwork in an aesthetic experience (cf. Heidegger, 2002, pp. 41–2; Hw, p. 54), but rather the allowing of oneself to be transformed by the work and to bring this transformation to a stand and to linger in it.[18] In the German word which we translate as preservation – *Bewahrung* – the word *wahr* (true) is clearly heard. Preservation is only then necessary for the work, if there is something in the work, which calls on the receiver to preserve it, and this is so only if truth is at work in the work. As such, the work *is* only through the preserver. The work can of course exist without a preserver, but then only as an objectively present thing. If truth happens in the work, if the work is the gift of unconcealment, then there must be a preserver. The relation between these is one of simultaneity: 'allowing the work be work [*das Werk ein Werk sein lassen*] is what we call its preservation.' (Heidegger, 2002, p. 40; Hw, p. 53) There is only a setting-in-work of truth when there is preservation. This means, as we will see, that the truth of the work does not lie primarily in its actuality, but rather in the possibility of engagement, which it allows.

The change in Heidegger's understanding of truth and the radicalization of 'transformation' is the precondition that allows for the thematization of preservation. If movement does not have its 'ground' in the order of Dasein, then not only must the possibility of preservation be placed within this movement, but it must also be that the principle of order of Dasein itself undergoes a shift. When the ontological difference itself is put into question, then the understanding of truth in *Being and Time* becomes problematic. Truth cannot simply mean the truth of praxis, because this would amount to forgetting the question of the truth of the difference of praxis and poiesis. The order of truth of this happening of difference can no longer be thought from Dasein, but rather as an order which sets Dasein into the possibility of truth. Heidegger thinks this order as the order of the work.

II The temporality of the work

II.1 Technē and the question of grounds

The question of historical action, and with it that of the relation between kairos and chronos, have been shown to be problems of truth. In Chapter 3, the question of truth was investigated in the context of issues of freedom and contingency. The question of freedom concerns action; to the extent to which action is performed kairologically, it goes beyond causal relations. In 'The Origin of the Work of Art', as we have seen, the work itself has no effect, and its making cannot be thought causally. What this suggests then is that the difference between kairos and chronos can no longer coincide with that between praxis and poiesis; making, in the sense at least in which the artwork is made, ruptures causal relations in that manner we have seen historical action described as doing.

In the 'Rector's Address', Heidegger translates the Greek word '*τέχνη*' with '*Wissen*', 'knowledge'. (Heidegger, 1985a, p. 472; R, p. 11) This is nothing new for Heidegger; he has on many occasions stressed that *τέχνη* is not so much an action as a form of knowledge. But now in 1933 *τέχνη* is understood as knowledge, which understands itself as historical, as an experience in which we suffer something from the having-been, in short *τέχνη* is understood as fate (*Schicksal*). (Heidegger, 1985a, p. 472; R, pp. 11–12) *Historical* knowledge is thus understood as knowledge as *grounding*. Furthermore, he goes on to etymologically interpret *ἐνέργεια* as 'setting-in-work' (cf. Heidegger, 1985a, p. 472; R, p. 12). Knowledge is being understood as grounding, and that which is grounded is being understood poietically as that which is set in work.

This reinterpretation is not so much a development of *Being and Time*, but rather marks a new understanding of one of the principle distinctions of that text; namely between possibility and actuality. One of the basic theses of *Being and Time*, as we have seen, is that possibility is higher than actuality. Actuality – *Wirklichkeit* – however, translates *ἐνέργεια*. In both the German and Greek words, we can hear 'work', in Greek '*εργόν*'. The actual is that which has been set in work. To the extent to which knowledge is understood as τέχνη, then the object of knowledge is the actuality of the work. However, the fact that the work is actual does not mean that it is objectively present. It is, of course, objectively present (the freight-handler or the charlady in the museum can confirm this[19]), but the being of the artwork does not consist in this objective presence. In the artwork, actuality is encountered not as objectively present, but as that-being.[20] According to the Artwork Essays, before the happening of the artwork a surprise befalls the one encountering it, the surprise namely '*that* this work *is* rather than is not' (Heidegger, 2002, p. 39; Hw, p. 51 [first emphasis mine]) The experience that there is something – an experience which, for Heidegger, lies at the basis of the 'why question' – is one which does not emerge for us faced with equipment,[21] but rather arises in the presence of the work.

The actuality of the work opens up the question of its possibility; it is the experience of this actuality which brings wonder with it. The work emerges from the collapse of the normal order of the objectively present, and 'the more essentially the work opens itself, the more luminous becomes the uniqueness of the fact that it is rather than it is not.' (Heidegger, 2002, p. 40; Hw, p. 52) The work is a grounding and in grounding the why question becomes possible: 'grounding something *makes possible the why-question in general*'. (Heidegger, 1998, p. 129; Wm, p, 166) The why question, however, asks not about the actual as such, but rather about the possibility of an actuality. Can this possibility be understood as something lying in the work itself, or does the work rather give the impetus to question beyond it?

In the wake of *Being and Time*, Heidegger increasingly questions the foundational role of Dasein. This requires at the same time a rethinking of praxis. This rethinking begins with a thematization of grounds. In *Being and Time*, Dasein is the supporting ground of the question of being (BT, p. 142; H, p. 152) and is the 'null ground of its nullity'. (BT, p. 283; H, p. 306) But what it means to be a ground is not discussed. Rather, Heidegger talks of 'null grounds' as if we already know what 'ground' is. Once we pose the question of grounds, then the relation of praxis and poiesis begins to appear much more fluid, especially as at the same time Heidegger is thinking through the implications of his account of historicity. The core question here becomes whether a historical action can be praxical without poiesis: a significant act, which is not brought to a stand, which is not 'set in work', is subject to dispersion and absorption in its effects, both intended and unintended.[22]

It is not immediately evident that Heidegger's essay 'On the Essence of Grounds' (1929) concerns either historicity or action. But the movement of this essay is one which proceeds from the static truth of the sentence to a pre-logical movement. Furthermore, in the essay, Heidegger characterizes transcendence, which forms its principal theme, as 'primordial history (*Urgeschichte*)'. (Heidegger, 1998, p. 123; Wm, p. 157) This term reflects an initial attempt to formulate what he will characterize in the Artwork Essays as a strife between world and earth. According to Heidegger in the 1929 essay, nature can only be manifest if it is made possible for entities to 'enter into' [*eingehen]* a world. This entrance into the world is an occurrence (*Geschehen*). Heidegger here still understands this occurrence on the basis of the transcendental structure of Dasein. The movement in question he sees as that of Dasein: through Dasein, transcendence as being-in-the-world transcends, something occurs amongst entities. We can hear in this text the interplay of light and dark, opening and closing, which will later dominate the Artwork Essays. The phenomenon of the work is already implicitly present in this text, something which announces itself also in the concept of establishing (*Stiftung*).

Heidegger does not pose the question of establishing – of grounding – in *Being and Time*, because Dasein is assumed as the ground. When, however, in 'On the Essence of Grounds' entities are disclosed only through the happening of entrance into the world, that means that that it is no longer through being disclosed, but rather in a specific and unique moment of 'giving of grounds', that entities come to appearance. Ground is no longer understood as praxis, it must rather be established. Grounding requires something in order to ground (found) something upon it. This something cannot itself be grounded, but rather comes to appearance in the grounding. That which comes to appearance is the entity which was not yet 'there' in its being. Grounding is a leap beyond – transcendence – and in this leap beyond darkness comes first to appearance. This darkness of entities is a resistance against transcendence, which is not dissolved in an entrance into the world, but rather must be continually opposed. Entities do not exhaust themselves in their worldliness; there is being only in transcendence, but that is not the same with entities: 'ground … belongs to the essence of being because being (not entities) is given only in transcendence as a grounding that finds itself in a projecting of world.' (Heidegger, 1998, p. 132; Wm, pp. 169–70) In entities, there is a darkness which can never be seen through. Only through a specific happening do entities come to appearance. But in this essay, the as yet unasked question is *how* this happening occurs.

Although this question is not posed, the answer is already implicit. The happening of entities is investigated in the context of poiesis and according to the paradigm of the work. The happening is itself a thrown projection. (cf. Heidegger, 1998, pp. 134–5; Wm, p. 172) The projection of understanding, in *Being and Time* still a possibility of praxis, is reinterpreted in this essay as a matter of establishing and can as such be understood as projection in the manner of an artist or an architect.[23] Thrownness is understood in the sense of *Being and Time* in relation to fallenness as being absorbed (*Eingenommenheit*) in entities, but the latter is now reinterpreted as 'taking up a ground or basis' (*Boden nehmen*). (cf. Heidegger, 1998, pp. 128; Wm, p. 164) In its thrownness, Dasein takes up a *specific* basis (*Boden*) in the midst of entities. Basis and world are not the same: world means opening, basis marks that which necessarily remains closed to all opening. If the basis is a specific one, it must be determined by the world, because only in the openness of world can difference, and hence specificity, emerge. In the work, the basis is in this sense determined and moulded. To the extent to which world is formed through the world, such a basis is formed into a home.

That the happening of the thrown projection fulfils itself in making seems both surprising and problematic in the context of *Being and Time*. In the latter work, history is the occurring in its specificity.[24] This means that the unique within which history exists is itself the occurrence. If now occurrence is brought together with grounding – so with poiesis – then it would appear that the decision between the praxical and the poietical character of history has already been made.

We should be clear what is at issue here. We have already traced the experience of kairos in relation to praxis. Now it appears that Heidegger is showing kairological traces in the work. However, there lies in poietical experience as we have discussed it, a strong tendency to chronological time. There must then be a change in the concept of poiesis as Heidegger is understanding it. That is indeed the case, as will become increasingly clear.

Every attempt at grounding ends in an abyss (*Ab-grund*). Such is the conclusion of 'On the Essence of Grounds'. The abyss of Dasein is, however, freedom. (cf. Heidegger, 1998, p. 134; Wm, p. 171) Freedom is the self-binding, which can only be justified subsequently: '*Freedom is the origin of the principle of grounds.*' (Heidegger, 1998, p. 132; Wm, p. 170) Freedom happens in the realm of possibility. It is the origin of grounding, but not grounding itself. In grounding, Dasein is referred to freedom, but freedom is not the cause of the work. The work cannot be explained in terms of what went before, but is rather with respect to the letting free, the un-concealment of being, to which the work is a response. Freedom is the essence of truth. When truth is set in work, the work is first made possible through the original, that is, free-releasing, temporality. This temporality is the time of truth, i.e. of the happening of the unconcealment of being. Truth is a possibility, which comes to appearance through the 'that-being' of the work, but is not exhausted in it. Truth is set in work, but it is primarily the possibility of this 'setting-in-work'. Grounding then does not form truth. This happens in a transformation, which is recognizable only in the phenomenon of being-brought-to-a-stand. This coming to a stand can, however, only happen through a grounding. Furthermore, the abyss comes to appearance only in the grounding, in the failure of grounding, the failure of the attempt to ground something.

In *Being and Time*, this coming to a stand is accomplished in the disturbance of the equipmental context. The movement is hence disclosed in its truth in the form of repose, but at the same time the aspect of kairological temporality is lost. To bring this temporality to a stand, what is required is a work which emerges out of the suddenness of the moment of vision. This formulation is, however, misleading. The coming to a stand forms an order in which Daesin can first recognize kairological time. In *Being and Time*, Heidegger believed he had found this order in Dasein as praxis. But now the question concerns the condition of possibility of this order, of its coming to a stand.

In 'On the Essence of Ground', praxis no longer forms the underlying order. Dasein is historical. Praxis is *one* form of historical action. At the same time, praxis is in danger of historical dispersal. Nevertheless, in praxis we find the possibility of action before or beyond all grounding. A freedom becomes apparent which reveals itself in its own contingency. This freedom lies at the basis of the 'why question'. If the human being was able to dispose of the possibility, to move within this freedom, this movement would form the original encounter of human beings with truth. Only to the extent to which the human being goes beyond grounds, can she question the ground of grounds and open herself to the kairological truth of the happening of being. That means, however, that in every movement in freedom a truth emerges, which is not set in work, but at the same time is only perceivable in the setting in work. The possibility of human action is placed between movement and repose. The work alone contains within itself the possibility of grounding, which is the forgetting of the abyss.

If the human being did not sense the lack which develops out of this forgetting there would be no desire for philosophy. This desire emerges because grounding always refers to an ungrounded, in which the abyss can be traced, an abyss which Heidegger later terms 'the earth'.

II.2 The having-been of the work

II.2.a The struggle between earth and world

As we have discussed, in *Being and Time*, Heidegger had seen in the work the possibility of access to the having been of a past historical world. It is for this reason not surprising that in understanding the historicity of Dasein in his later work, Heidegger turned to the being of the work in opposition to that of equipment. It is important in this context first to establish what the 'setting-in-work' means, as an approach to the problematic of the having been.

The setting in work of truth arises out of the interplay of possibility and actuality, which Heidegger characterizes as that of earth and world. Movement is brought to a stand in the work and only thus can there be truth. If this 'coming to a stand' means actuality then it might seem as if there is truth, or at least that truth can only be validated, in the realm of actuality. In *Being and Time*, however, Dasein as possibility, not Dasein as actuality, is the 'place' of truth. In the work something is grounded. The setting in work is a transformation, in which 'in its exclusive actuality [*Wirklichkeit*], what went before is refuted.' (Heidegger, 2002, p. 47; Hw, p. 61)

In his influential book, *Möglichkeit und Wirklichkeit bei Martin Heidegger [Possibility and Actuality in Martin Heidegger]*, Wolfgang Müller-Lauter attempts to show that the priority of possibility and the connected priority of futuricity in *Being and Time* does not withstand critical examination. According to his analysis, each modality corresponds to an ecstasis of temporality: possibility corresponds to futuricity, necessity to the having-been and actuality to the present. (Müller-Lauter, 1960, p. 28n. 11) The priority of the future though, is in such a view not shown by Heidegger since his analysis assumes throughout the actuality of death. As he puts it, 'Death must already have been made temporal as an event, for Dasein to anticipate it as its ultimate *futural* possibility. That means: the future arises out of the original present of death.' (Müller-Lauter, 1960, p. 53) This judgement corresponds essentially to the conclusion reached in Chapter 2, although there death is understood in terms of future past rather than the present. However, on this basis, Müller-Lauter seeks to go further and to argue for the priority of actuality in discourse (in relation to conscience) and the priority of possibility in fallenness. What he ignores in this account is the difference between authentic and inauthentic temporality. Precisely because Dasein is not simply an observer of time, but rather is understood as an agent in and through time, each of the modalities can appear in each temporal ecstasis. Contingency as well as necessity concern the whole temporalizing of time, as was argued in the last chapter. This can be seen above all in the case of the present.

The problem of the role of the present in *Being and Time* cannot be resolved by equating the present with actuality, as Müller-Lauter does. Actuality itself remains unclear, if it is understood in terms of the temporalizing of temporality. With respect to such temporalizing, so in relation to the temporality of Dasein, the analysis of *Being and Time* stands or falls on the priority of possibility, a priority based on the practical constitution of temporality. The present is indeed the temporal ecstasis of the being-together-with (*Sein bei*) the objectively present; but equally in the authentic present of the moment of vision Dasein is disclosed as possibility. Even if it were true, as Müller-Lauter claims, that conscience 'is primarily grounded in a presenting'. (Müller-Lauter, 1960, p. 63) This would not mean that discourse has the character of actuality, as he claims (cf. Müller-Lauter, 1960, p. 64), because it can have the form of a call (the call of conscience) in the moment of factical action.

Müller-Lauter's analysis of the Artwork Essays betrays the same mistake. He develops an account of the present that ends in a timelessness which he sees as characteristic of the artwork. This timelessness consists in the actuality of the work. Heidegger, according to Müller-Lauter, did not see this because he subordinated the actuality of the work to its possibility and therefore could only understand the work in relation to the 'letting happen of truth as the opening of the free space of world and history.' (Müller-Lauter, 1960, p. 87) It is, however, in no way clear that Heidegger does place possibility higher than actuality in the Artwork Essay. In fact, the more obvious reading is the opposite – Müller-Lauter underestimates the shift which has taken place since *Being and Time*, and the growing emphasis on the work in Heidegger's thought. As the problem of grounds comes increasingly to the centre of his thought, the question of possibility becomes a problem in relation to the work more than to Dasein. Müller-Lauter does not see this clearly enough, and thus can – in a manner consistent with his overall interpretation – understand the 'that (Dass)' of the work not as residing in its uniqueness (as I will attempt to show in what follows), but rather in its being made, and hence in its relation to the possibilities of Dasein. (Müller-Lauter, 1960, p. 91) The issue of the Artwork Essay, as I will attempt to show, is that the action of creating has to be understood on the basis of the work. That the problem of action is of fundamental importance here is not acknowledged by Müller-Lauter. Action always takes place in the realm of possibility, that is – in this context – that possibility opened up by the strife of world and earth. What Müller-Lauter fails to see is the moment that is constitutive of all authentic relations to the work: that the actual opens up the possibility of these relations.

Only on the basis of modality can we understand Heidegger's claim that 'the work-being of the work consists in strife as contesting [*Bestreitung des Streifes*] between world and earth.' (Heidegger, 2002, p. 27; Hw, p. 35 [translation modified]) The work is the that-being of this strife. In its contingency, in the possibility of its non-being, thus in opening up the question of its being as possibility, the actuality of this that-being is not objective presence, but the 'thrust into the open' of the uniqueness of the work. (Heidegger, 2002, p. 40; Hw, p. 52) The uniqueness of the work lies at the basis of its been–made, not the other way around, as Müller-Lauter claims. (Müller-Lauter, 1960, p. 91) That 'in the work createdness is expressly created into what is created' (Heidegger, 2002, p. 39; Hw, p. 51; quoted in Müller-Lauter, 1960, p. 91) does not lie in the power of the maker himself. The artist remains 'something inconsequential in comparison with the work'. (Heidegger, 2002, p. 19; Hw, p. 25) The act of making indicates something which first comes to appearance in the work, namely the uniqueness of its being. As Heidegger puts it:

> the work casts before itself the eventful fact that, as a work, this work is and exhibits this fact constantly. The more essentially the work opens itself, the more luminous becomes the uniqueness of the fact that it is rather than is not. (Heidegger, 2002, p. 40; Hw, p. 52 [translation modified])

Only if this uniqueness happens as an eventful fact can we speak of the specific that-being of the work. But uniqueness indicates that actuality in a certain sense changes over of itself to possibility. This is so because the unique *is* only in repetition, which takes place not in the realm of actuality, but in the free play of possibility. The work is not 'in itself' unique, but only because in itself it brings a transformation to a stand, a transformation which only shows itself as possibility within it.

Although 'world' plays an important role in *Being and Time*, 'earth' hardly appears in this text.[25] Earth has no place in a fundamental ontology of Dasein, nor can it be explained on that basis. Nature appears in *Being and Time* only under the category of use. (cf. Hüni, 1993, p. 299) Earth appears as the material of human action, yet even as such seems to point to a dimension beyond will and intention. The question of grounds as it opens towards the absence of ground, the *Ab-grund* (abyss), makes the question of that in human action which escapes will and intention of increasing importance. In the piece of equipment, earth does not appear; it is rather levelled off into 'stuff', and is exhausted in serviceability. In handiness, the abyssmalness of the earth is covered over in the purposiveness of equipment. The past of the equipment appears only when it does not function; at work, the equipment disappears into the future, in its directedness towards a goal. The work, however, sets (-duce) something forth (pro-), producing – *her-stellend*. That from which a work is produced is 'not stuff as a means for the manufacture [*Verfertigung*] of something' (Heidegger, 1989a, p. 11), because it does not let the material disappear in a directedness on a goal, but rather lets it first come to appearance.

But that which comes to appearance is precisely a self-secluding (*Sichverschliessen*). (cf. Heidegger, 2002, p. 25; Hw, p. 33) This self-secluding shows itself in the actuality of the work. To show itself, the earth needs the illuminated opening (*Lichtung*) of the world. The work sets up (*Aufstellen*) the world. In the setting up of world there is also a breaking off. That which breaks off, which comes suddenly to stand, is not the unfolding of that which already was, but arises in opposition to the past. The world is not an entity, but rather the opening itself. The work does not enter an already existing opening, but rather it 'holds open the open of a world'. (Heidegger, 2002, p. 22; Hw, p. 30) The world 'worlds' in the work. (cf. Heidegger, 2002, p. 23; Hw, p. 30; also Heidegger, 1989a, p. 9) The world, according to Heidegger, is not a mere setting (*Rahmen*) for things; rather, as worlding, the world gathers spaciousness. There is possibility through this 'spaciousness' (*Geräumigkeit*), through the illuminated opening of world and, through such possibility, the self-secluding of earth comes to appearance for the first time.

What is of concern for us here is the having-been of the work which interplays in the strife between earth and world. The having-been of the work does not originally appear discursively. It is not so, according to Heidegger, that we understand a work through reports on the epoch to which it belongs, rather we understand the epoch, and hence the past world, through the work. The having-been of the work can be traced in the work itself. There is a lack of presence in the work, which is rather an excess of its presence, an excess which can never be fulfilled. As such the excess appears as lack: the work promises more than it can redeem. That means: the work opens a future, which does not have its origin in the work because the work can never fully 'produce' the earth, but rather remains 'imprisoned' in it. It is only because of this lack in the work, which announces an absence, that we can sense the having-been of this work and having-beenness overall.

To understand how this having-been is experienced, we can follow a hint from Heidegger, 'in the proximity of the work we were suddenly (*jäh*) somewhere other than we are usually accustomed to be.' (Heidegger, 2002,

p. 15; Hw, p. 20) In such proximity we experience a movement. The work does not move, it remains where it is; we do not move ourselves. Nevertheless, we are elsewhere although we are physically there in front of the work all the time. When, for example, we say, 'I was deeply moved by the music', movement is being understood here affectively. However, movement is not being used metaphorically. In the moment we are changed, have become other than we were, but in a state which was not caused by something previously present. We are not for that reason other than we were because we experienced an effect from many objectively present possibilities, which were contained in a cause. To become other means to be in a new state; the whole world becomes other for us. Understood ontologically, a metamorphosis has occurred in which the having-been is present, but has become strange.[26]

Heidegger hints at this strangeness with phrases such as 'elsewhere' (*anderswo*) and 'other that it was' (*anders wie sonst*).[27] The strangeness of the having-been consists in a double movement of withdrawal and 'essencing forth'. The phenomenon of metamorphosis is appropriate to illustrate this double movement because in metamorphosis the past 'essences forth' into the present, but in a form which opens up possibilities for the present, which would have been unthinkable in the past. 'Metamorphosis' is being understood here in its ancient, 'mythological', rather than modern biological meaning. In ancient myths, that which was becomes other, what it becomes is such that it retains the qualities that it had, but in such a manner that they are no longer recognizable. Hegel describes the result of transformation in relation to the myth of Niobe in this way: 'The rocks are no longer stone, but rather Niobe, who weeps for her children.' (Hegel, 1970, p. 505)[28] The having-been withdraws through a rupture, the moment of vision, in which it appears in a new light; at the same time, the having-been essences forth anew, as if for the first time, into the present. And its familiarity can only be glimpsed, and then only on occasion, in its present strangeness. A continuity remains, but an obscure one.

In our present context, metamorphosis does not refer to specific entities, but rather to the *order* of entities. The past order does not withdraw entirely. Face to face with the work, we find ourselves in a changed order, but this lets it be that we – to employ Heidegger's example – glimpse a previous everyday order of a peasant woman in its truth. The peasant shoes in van Gogh's picture are no longer familiar to us, they have becomes strange. But in their strangeness a world opens up. This world has-been. Precisely due to the levelling work of tradition, this world is familiar to us, its truth is lost. Only as a strange world does the having-been essence forth into the present. The having been is transformed in the present and is so such that new possibilities are opened up and old ones secluded off. This secluding off needs to be taken account of, because due to its metamorphosis is generally bound up with a mood of sorrow – sorrow, as we will see, is for Heidegger the mood of the poet.

Mood is indirectly referred to in the line cited above – 'in the proximity of the work we are suddenly somewhere other …'. It is not an accident that both here, and in the sentence referred to previously ('I was deeply moved by the music'), the past tense is employed. What this indicates is that the movement which is spoken of in both these cases is that which can only be determined *subsequently*. If someone says 'I *am* very moved by the music', then their words lack genuineness. We cannot at once be moved and describe ourselves being moved. This is so because the movement which occurs in the face of the work is one of mood and mood is speechless. (When I say 'I am sad' this is not a statement of sorrow, but rather a reflective determining of a given state.) The mood is speechless because it refers not to anything which has presence, but rather immediately to the having-been, which Heidegger interpreted as the 'that' of thrownness. The 'that' names the uniqueness of thrownness, which is always absent because it only happens suddenly.

When this movement in the direction of the strange occurs through mood, and mood forms the atmosphere of the situation, then we must assume that the work itself, which forms the situation, has a moodful character, indeed is a source of mood. If this is so, then it would appear that the moodful experience 'in the proximity of the work' opens up an approach to the relation of the work to its having-been. We sense the movement of the work in ourselves being moved.

As 'preservers' of the work or as thinkers we take part in grounding. The movement which can be sensed in the encounter with the work is one that comes to appearance only in grounding. Grounding refers to the having-been of the work as the *ground of grounds*. What then is the ground of grounds, why is ground given? The question aims not towards some psychological motivation, but rather to the inescapability of grounding in human action. It cannot be answered simply by reference to making, as it is precisely the basis of making which is here in question. It is a question which can only be posed subsequently. As such it concerns the having-been. In encountering the artwork, this question cannot be answered with a causal explanation, because the artwork marks precisely a break with the normal order. As Heidegger puts it:

> the more purely the work is itself transported into the openness of entities it itself opens up, then the more simply does it carry us into this openeness and,

> at the same time, out of the realm of the usual. To submit to this displacement means: to transform all familiar relations to world and earth. (Heidegger, 2002, p. 40; Hw, pp. 52–3)

Our question aims not at causal relations, but rather at freedom; it is a question not – despite appearances – about the presence of grounds, but about their absence (*Ab-grund).* The earth is abysmal. We hit here upon an inner connection of freedom, the having-been and earth.

Grounding encounters the abyss, which limits its possibility. The having-been is never entirely gathered together in grounding. This is not to make the trivial point that grounding must always make decisions regarding the having-been, but rather more substantively that grounding always comes up against the resistance of the having-been. This resistance of the earth lies within the structure of truth.

The movement of truth as Heidegger describes it cannot be thought without reference to freedom. Freedom is a movement which lets us reach back from a work, thing or statement to the original happening of truth. Freedom is at the same time the breaking open of the abyss in the grounding transcendence. (cf. Heidegger, 1998, p. 134; Wm, p. 172) In freedom, an entity comes into the light of the world; at the same time, this coming to light appears as abysmal. For this reason, Heidegger can say that the truth *is only* in the work (Heidegger, 1987, p. 191; EiM, p. 146), because, in grounding, the freedom of transformation appears in the coming to a stand, in the work, as does the being imprisoned in the abysmalness of this transformation. The occurrence of truth consists, however, not simply in the appearance of something, but also in the un*concealment* itself. Concealment remains inconspicuous in the everyday encounters with things, indeed is itself concealed. (cf. Heidegger, 1998, p. 148; Wm, p. 191)

It is only when the question concerning being, the why-question, arises that the 'enigma' of truth appears. This happens in the 'that-being' of the work.

The 'that-being' of the work involves both time and place: that it was created in a certain place and at a certain time. This time and place are both brought to appearance in the work. However, the place is more than that which is founded in the work and the time is not exhaustively conditioned by the work. There is a having-beenness that cannot be gathered into the work and a spatiality which does not appear in the light of the world. That the latter can be termed 'earth' makes a certain intuitive sense. However, in what sense is the time, which as having-beenness is collected in the work and yet resists it, also earth? To answer this question we need to investigate the earthiness of the work.

Stated briefly, the ground of the work is the abyss (*Ab-grund*). This claims at least two things: firstly, that that on which the work is grounded is itself without ground; but secondly, that immanent to the work, in the ground of its that-being, there is something abysmal. Where this abyssmalness is to be traced is hinted at by Heidegger when he stresses that material is used up (*verbraucht*) in equipment, but is made use (*gebraucht*) of in the work, such that the material first shows itself in such use. (cf. Heidegger 2002, p. 25; Hw, p. 33) The work is so constituted to bring the earth to appearance in this manner. At the same time, an opening of world occurs in the work. Nevertheless, earth resists world.[29] This resistance does not simply appear in the work, it constitutes the work. World happens in the strife against this resistance, but the ground of the work does not lie in the world, but rather in the earth. The double meaning of abyss (*Abgrund*) is reflected in an ambiguity in the way in which 'earth' is employed by Heidegger in the Artwork Essay. On the one hand, earth means the 'coming-forth-concealing', which, for example, characterizes the heaviness of the stone in a sculpture; on the other hand, it is that within and upon which human beings dwell in the world. (Heidegger, 2002, p. 24; Hw, p. 24) It is not immediately clear how these two meanings of earth are connected.

As that in and upon which it dwells, the earth is that into which the human being is thrown. The earth embodies the historical having-been of such a being. The history of a people is at the same time the history of its territory (cf. Held, 1991a, pp. 333–6), which through the work is moulded into a home: 'the earth becomes first worldly through the work and as such is turned into home [*Heimat*]'. (Heidegger, 1989a, pp. 12–13) Understood in this way, earth is that on which human beings take up place. This history finds a form and a shape in the work. But the work does not simply transmit that history; rather the having-been remains hidden in it. In a Medieval tapestry or in a Baroque church a secret is hidden which can be characterized as the strangeness of the having-been. The work secludes itself, it does not emerge fully in the present. It remains having-been and withdraws from the present; it moves away from the present. But only through this movement can there be a 'within', a place which is not exhausted in the grounding of a dwelling. This movement-into-the-having-been binds together the earth as self-secluding, and as that within and upon which humans dwell. Without this movement there would be no human dwelling, because if the having-been would merge completely with the present there would be no place, only an abstract space in which everything would be equally and indifferently available to a neutral observer. A dwelling – similar to, say, a sculpture – can only then be unique, that is historical, if in the present a lack – that which cannot be made available – is traceable.

Both in the artwork and with his dwelling, the human being encounters the phenomenon of earth. In the abyssmalness of earth, he experiences the fragility of time, which cannot be reckoned with, but which can be 'grounded' only in the experience of the failure of grounds. It is in that failure of grounds that the possibility of a new foundation, of a founding and instituting anew – what Heidegger refers to as *Stiftung* – emerges. As he puts it in the first Hölderlin lecture course in the winter semester of 1934–5, '[I]t is always the case that the great times of change (*Wendezeiten*) of a people arise out of the abyss (*Abgrund*) and in each case to the extent that as a people reaches down into it and that means into its earth and occupies its home.' (Heidegger, 1989, p. 106) The earth has a priority over world because in it the kairos is experienced – experienced, that is, in grounding and founding of the artwork and the dwelling.

This priority of the earth can be also found in the structure of truth. According to Heidegger, the concealment of entities as a whole is 'older as each openedness of this or that entity'. (Heidegger, 1998, p. 148; Wm, p. 191 [translation modified]) 'Older' must be heard here kairologically. It would be senseless to say that concealment occurred 'before' the letting be; rather, what is meant is that concealment shows itself in an unconcealment, which cannot come into the present, but which is traceable as a withdrawing having-been in the present. The abyssmalness of freedom shows itself here in the final analysis as a relation of freedom which transcends itself. This is so because, in the letting be, we sense the abyss of freedom, although this is not to be found in freedom itself, but rather in the earth.

Earth needs the world. 'Earth cannot do without the openness of world if it is to appear in the liberating surge of its self-closedness.' (Heidegger, 2002, p. 27; Hw, p. 35). The movement of earth is one of return to itself in opposition to the self-opening of the world as a flowing into the future. But this movement of the earth only appears through the strife between world and earth, which is founded by the work. This movement is nothing present, but shows itself in the movement of return into the absence of the having-been. This does not mean that this movement is one of passing away. The earth is nothing objectively present which passes; it essences through the work into the future. In a fundamental way, the human being experiences the earth both in the artwork and in its dwelling in the abyssmalness of its thrownness, i.e. of its relation to a having-been, which lies outside the reach of its power of disposal. That having-been, though, is at the same time that to which the human being must always call upon for the *justification* its action.

With this reference to justification, we can get a little closer to answering the question concerning the ground of grounds. To justify something is to find a ground in the past for it. The future does not contain any justifications. Future consequences can of course be appealed to, but these are grounds for the action, not in terms of their future occurrence, but rather as they moulded the agent's understanding of the past world out of which the action emerged. Justification always takes place subsequently, but it seeks grounds in the past of the action, not in its future. It is in this attempt to find grounds in the past that the abyssmalness of action appears. The structure of justification brings to light a tension between action and reflection, which is not to be found in Heidegger's analysis. He states in a lecture course from summer semester 1934: 'we should delve into action and come out of reflection [*Wir sollen ins Handeln, aus der Reflexion heraus*]'. (Heidegger, *Logica*, p. 14; cf. GA 38, p. 53), as if action was possible outside of any relation with reflection. If, as Heidegger tells us in *Being and Time*, the '"one" takes from Dasein its own responsibility' (BT, p. 119; H, p. 127), this also means that it takes over from Dasein all justification. Inauthentic Dasein is never justified, because its actions do not seek their own justification, but rather imitate what 'one' does. The striving for justification is a striving for a ground. But the ground is only present in grounding, that is in action. Nevertheless, reflection is a coming back to the past in order to justify an action. This assumes that the grounds for the action are to be found in the past. It also assumes that there is a present which is still, and which gives the human being the place for such reflection. In Heidegger's analysis of the present – both in *Being and Time* and in the writings throughout our period (that is to 1936) – there is no recognition of the quiet and still present of reflection and contemplation. In the 'Letter on "Humanism"', which in many respects sums up much of Heidegger's thought in the aftermath of *Being and Time*, Heidegger says, '[T]hinking acts insofar it thinks.' (Heidegger, 1998, p. 239; Wm, p. 311) Action, it appears, has 'no time' for reflection; it has no access to a quiet and still present. Both inauthentic time and the time of the kairos are – albeit in different ways – times which are directed towards the future. But reflection is not something external to time. Rather, it is a moment of non-action within action itself. We can see this when we notice that in repetition, which characterizes authentic historical action, the question necessarily arises as to *what* should be repeated. Not every repetition can be justified; if that were not the case there would be no critical distance to the past. But Heidegger's whole project of destructuring the history of ontology is one of creating such a critical distance. Despite the fact that Heidegger has been continually criticized for lacking such a critical distance, in fact this distance lies at the core of his thought, in particular in the distinction between kairos and chronos. The difficulty arises here, however, because reflection is, for Heidegger, characteristically chronological. What occurs in reflection is not a repetition of the past, but rather the attempt to order it into a chronology. Such chronological modes of justification operate by making recourse to chronological continuity.

In the kairological situation, however, there is not time for reflection. It is only after the event that the chronological significance of an action begins to emerge. Chronology has the tendency towards unity and completeness, and although the kairological situation is experienced as a rupture, it is subsequently integrated into a renewed chronology. The desire for justification motivates this tendency towards completeness. Such completeness is, in the end, one which is testified towards an other. The human being justifies herself before the other, who functions as a judge. Such attempts at justification assume responsibility. As Heidegger puts it in the same lecture course just quoted: '[A]lso the answer to the question concerning history has the character of decision. … in responding we take over the ordering of the occurrence and make it into history. *This responding is a taking of responsibility*.' (Heidegger, 1991, p. 76, cf. the corresponding passage in GA 38, p. 121)[30] The occurrence does not occur except in the response; in just the same sense, Heidegger says that the work of art is not except through preservation. The question *is only* through the possibility of response; the artwork *is not without* the possibility of preservation. The response and preservation form the future. Understood chronologically, the future is only that of responding and preserving human beings. Within chronologies – in the chronological striving for unity – questions are either answered or not (yet). The form of questions and responses which are Heidegger's concern here, 'are such in themselves, that they are never settled' (Heidegger, 2009, p. 101; GA 38, p. 112) This is so because the questions concern not this or that, but rather the very order of entities itself.

Through our response, we transform the mere chronological event into history. We do this to the extent that, in responding, we take over the consequences of occurrences. Within the kairological situation, the occurrence is taken up in responding to it. It is only subsequently, in the establishment of a new order, that the question posed can be settled. Once the human being – a human community, in fact – comes to terms with such an occurrence, then the response becomes an answer which can be taken to have settled the matter. In such a coming to terms, the occurrence is allowed to slip into the past and become a part of a chronology. Indeed, without such a slipping away of the occurrence there can be no history. The work in its actuality – understood in terms of its actuality – becomes an answer which admittedly can never fully take up responsibility for the occurrence itself.

In justification, chronos and kairos cross over one another, and this crossing over makes history possible. The seeking for justification is itself a taking over of responsibility. Such justification takes place before an other. The new order of entities is taken over as a step beyond the past, whether or not it is taken as progressing on that past. As such, the responding and preserving, essential as they are for the kairos, cannot be understood without reference to chronos. What this means is that the full force of the question and the full force of the work as an emergence in the moment of vision fall necessarily into forgetfulness as the response and the preservation take the form of a settling and a coming to terms. But this is precisely essential for history itself.

The work in responding to the earth in the materiality of its createdness assumes the space between world and earth. If, however, the transformation comes to presence in the being brought to a stand, then the space is not a space but rather a rift (*Riß*) between actuality and possibility, without which there would be no art and no artworks. (cf. Heidegger, 2002, p. 38; Hw, p. 50) This rift is at the basis of the paradoxical relation of the work and the 'things of the earth'. The paradox lies in the fact that the work is a rejoinder to the art-like in nature, but at the same time such art-like is only opened up 'because it [nature] is lodged in the work in a primordial way.' (cf. Heidegger, 2002, p. 38; Hw, p. 50 [translation modified]) That which originally is lodged in the work comes from an origin, which can be traced only in the work. As such, possibility and actuality are not two regions, but rather possibility appears in the actual and in the possible we encounter actuality. Only in the rift in entities – that is, only in the difference and crossing over of possibility and actuality – can there be art and artworks. In the violence of this rift there occurs not only the difference between world and earth, but also the origin of the artwork as such: art.

II.2.b Poetry, art, origins

The work is a response to the having-been, which claims human beings.[31] The question of origins relates directly to this claim. According to Heidegger, art is the origin of the artwork. Art is, in this sense, the has-been. The work responds to art as has-been. The having been of the work cannot be brought to presence in it; there is a leap between the work and its past. In this sense, that which has-been cannot be given a chronological order. There remains in every chronology of a work that which exceeds that chronology because the having-been can never be fully absorbed into it. The origin (*Ursprung*) of the work only emerges in its attempt to return to the inexhaustible having-been. The beginning (*Anfang*) always happens abruptly, suddenly. As Heidegger states: 'Bestowal and grounding [*Schenkung und Gründung*) have in themselves the abruptness [*Unvermittelte*] of what we call a beginning.' (Heidegger, 2002, p. 48; Hw, p. 62) Thanks to the beginning manifest in the work, the situation of the work is always new; and thanks to this, the human being within the situation of the work is capable and positioned to act otherwise than previously. Action within this situation is the constant attempt to reach back to the origin disclosed in the work. This produces a tension within action without which it would remain inauthentic.[32] This tension characterizes the kairological situation in which the tension between future, having-been and present compresses itself into the moment of vision.

The striving to return to origins is a seeking for justification, an attempt to work against the threat of dispersal, in order to remain in the power of the beginning, which itself neither requires nor grants justification. Justification is, in Kant's terms, 'deduction'.[33] Is the relation of the artwork to art for Heidegger then something akin to a transcendental deduction? In the lecture courses which Heidegger held around the time of the publication of *Being and Time*, he stresses the temporality of the a priori many times and interprets it both poietically and chronologically. In the case of the artwork, art is 'prior' to the work, but not chronologically earlier. What is at issue here is not an order of right as opposed to an order of fact, but rather the event of order itself within entities as a whole. There is no kairological occurrence without a response related to it. In this sense, art and the artwork are contemporaneous. As such, the 'priority' of art as origin can only be understood kairologically. That means that we can sense the having-been of the work in the simultaneous occurrence of art and work. In contrast to a transcendental deduction, there is here no necessity in this justification. The artwork responds to art as to its origin, but how it responds, to that we can find no ground in art. The relation of artwork to art is a contingent one. And this contingency makes it impossible to speak definitively of a transcendental deduction of art and artwork. Nevertheless, the act of grounding is conditioned by the work, and the work itself displays a striving for the origin. The origin happens in the rift mentioned earlier. In and through a rift of possibility and actuality, a transformation is brought to a stand in the work. In this transformation, the origin can be traced. But this origin cannot be found in the work; it is not visible in it. It can only be 'heard' in the work. The one who can hear, Heidegger tells us, is the poet.

It is not by accident that Heidegger constantly stresses the fundamental mood of the poet, for mood relates to the having-been. The poet encounters the having-been in a fundamental mood. Heidegger – following Hölderlin – terms the founding response of the poet to the having-been Remembrance [*Andenken*]. Remembrance is not simply a remembering of that which has happened, but rather a response to that in the having-been, which makes a claim on the poet. (cf. Heidegger, 1996, p. 80) To express what it means to have a claim made upon the poet, Heidegger takes up a line from Hölderlin's poem '*Germanien*': 'as you begin, so will you remain'. (cf. Heidegger, 1989, p. 241) That which remains is that which begins, the origin. We do not find that beginning or origin chronologically. The origin is not the first step – indeed it is not a step at all – but rather a leap, that is, a sudden emerging of movement, a beginning out of discontinuity. Such discontinuity is at the same time the arising of the demi-gods. As we have already seen, demi-gods indicate transformation. They do not leave the origin behind them, but rather – to use the example of the demi-god 'the Rhine' from Hölderlin's eponymous poem – the whole stretch of the Rhine belongs to the origin.[34] The origin is sensed in the transformation itself. Understood as metamorphosis, transformation lets the having-been come to expression, and precisely as that which is strange.

The origin is strange and alien to the human being, and can fail to be heard in the everyday. In his first lecture course on Hölderlin, Heidegger understands the relation of hearing and failing to hear the origins (cf. Heidegger, 1989, p. 200), which clearly reflects the distinction between authenticity and inauthenticity in *Being and Time*, in terms of the difference of mortals and gods. The gods hear compassionately, and this merciful hearing Heidegger terms responsive hearing – *Erhören*; mortals hear as not being capable of hearing, and such hearing is a failure to hear or a wishing to fail to hear – *Überhören und Überhörenwollen*.[35] Mortals flee from the origin. The poet, however, plays a mediating role. As poet, 'he cannot wish to fail to hear the origin'. (Heidegger, 1989, p. 201) The relation of the poet to origins Heidegger expresses as follows: '[B]ecause poetry as founding of be-ing [*Seyn*] is the same origin as that which it genuinely founds, for that reason and only for that reason it is possible for poetry to speak of be-ing, it must even do this.' (Heidegger, 1989, p. 252) The poet stands between mortals and gods, and poetry has its origin in this between space. This space is only as transformation, and poetry is only as responding to such transformation. The origin cannot then be seen, it is not possible for it to be brought narratively to language, rather it can only be heard. It is only in the language of poetry, which is more of an 'enveloping saying [*verhüllendes Sagen*] than a progressive narrating and describing.' It is a saying which comes out of a 'harkening knowledge concerning the originating origin in its flowing forth [*Entspringen*]' (Heidegger, 1989, p. 203) In striving to return to the origin, the poet perceives it only in its echo; he finds it in the traces of the force of its beginnings in the sudden emerging of movement. The poet senses this force through mood, and only in it can this beginning of movement of the kairos be experienced. For this reason, Heidegger calls mood an 'originally unique motion [*ureigene Bewegtheit*]'. It is the motion of the 'having-been, i.e. the still essencing [*Gewesenen, d.h. noch Wesenenden*].' (Heidegger, 1989, p. 107)

The force of the origin is hidden in the everyday. In *Being and Time*, Heidegger considered that the everyday held within itself the possibility for this force to be sensed; every situation, no matter how ordinary, could be the occasion for Angst. (cf. BT, p. 177; H, p. 189) In the 1930's, Heidegger's account of mood shifts and this is related to the fact that the occasion for mood in the deep sense – so fundamental moods (*Grundstimmung*) – is no longer understood to be found in the everyday.[36] The force of the origin is not to be found in the normal experience of things. This is so because the relation with equipment, which lies at the basis of such normal experience of things, does not offer any original access to historicity. Only in the work is there that lack or (and these are inseparable) excess which gives occasion for the moods of sorrow (*Trauer*) or wonder (*Staunen*).

So understood, the origin is not the source of the last justification, but rather is the force of the emerging of movement. If these two were the same, then we would come close to a type of Christian Aristotelianism in which the human being finds his final justification in God as creator. What is of concern here is not a causal relation – understood in modern or in Aristotelian senses – but rather, understood kairologically, the abysmal movement of being and the poetic rejoinder to it. The question of justification leads not to grounds, but rather to the lack of grounds of the origin, which is itself not a cause. The possibility of the work lies outside chronology because it is not exhausted in the work and remains inappropriable for human poiesis.

The poet listens to the voice of the having-been and hears the origin to the extent to which she harkens to it.[37] The hearing of the poet is a hearing of that what is not yet set in the work. This raises the question as to the actuality of poetry itself and indeed parallel to the relation of art and artwork; Heidegger understands poetry as the origin of the poem – the linguistic work. This distinction has an ontological and a methodological significance. Ontologically, poetry is not a 'said' but rather a 'saying'. Methodologically, Heidegger is consistent with this to the extent to which he attempts not so much to analyze the said – the poem – but rather to 'say' the poetizing itself, that is, to repeat it and as such to set it in motion again. That which stands on the paper is not poetry, at most it is the occasion to poetry, which is always a saying. (cf. Heidegger, 1989, pp. 41–2, 45–6; on this theme, cf. Ziegler, 1991, pp. 27–8) How then does poetry and the poem relate to one another? Heidegger repeatedly cites Hölderlin's words, 'but what is lasting / that the poets found [*was bleibt aber / stiftet die Dichter*]'. ('Remembrance', Hölderlin, 1986, p. 211) The problem of poetry is founding. Founding for Heidegger means two things: firstly, to project essentially that which is not yet; and secondly, 'to keep safe that which is already said and grounded ... to save it as remaining remembrance for the essence of be-ing, which it opens up, to the remembrance of which a people must always turn their thoughts anew.' (Heidegger, 1989, p. 214) Grounding and remembrance belong together for a historical being, because grounding is guided by knowledge that 'tidings [*Kunde*] of it can be preserved'. (Heidegger, 2009, p. 75; GA 38, p. 87 [translation modified]) There can be preservation only because there is transformation, because what is to be preserved is precisely that which is founded in response to origins; the poet founds such transformation as she founds being. Founding is projecting. The poet projects being. Such projection is a response to being. Being is first a historical occurrence when the poet projects it. The founding of being is a grounding in the language of a people, which gives tidings of being.

Poetry is the attempt, to bring the uniqueness of the occurrence of being to speech, 'the poet always speaks as though the entity were being expressed and invoked for the first time.' (Heidegger, 1987, p. 26; EiM, p. 20). Poetry is, as such, historical in the sense that it belongs to a particular people, because it attempts to ground the distinctiveness of this people on the basis of the uniqueness of its experience of the occurrence of being. This attempt is practically impossible, and for that reason poetry is always in danger of falling back into prose.[38] The tendency in language to treat the unique as a particular, and as such to subordinate it to the general, means a levelling off of language.[39] In poetry, on the other hand, language is governed by fundamental moods. The speechlessness of mood in no way means that mood lacks a relation to language. In fundamental moods, human beings encounter the unique, and the speechlessness of mood arises out of the striving to bring this experience to language. The having-been is accessible in its uniqueness only through mood. This uniqueness makes a claim on the poet, to which she responds in language. 'Language' here does not designate a generic concept, but rather an occurrence, which happens in one go. As language binds together in conversation (*Ge-spräch*), being happens. That means that language does not come from human beings, but happens to them, to the extent to which it claims them. When we speak we never genuinely begin in speaking; rather language responds to something else.[40]

Jacques Taminiaux (Taminiaux, 1991, p. 210) did not take sufficient account of this moment of response when he interpreted conversation in the first Hölderlin lecture course in line with *Being and Time* as the conversation in conscience of the self with itself (an interpretation of conscience which also is not unproblematic, if the line of interpretation pursued in Chapter One above is valid). It is rather the case that, in conversation, the gods address the human. Language allows the human being to sense that he is not the final order of things, and not even the first. When we speak, we always come too late. This too late is that which marks the pathos of the speaking in relation to the flown gods. Human language responds to the gods, but the gods are absent. This does not mean (for Heidegger, though perhaps it does for Hölderlin) that the gods were once present in a golden age and have since flown, because that would be to illegitimately (and indeed mythologically) chronologize the gods. As Heidegger says, 'such a present of the gods has been. However, if we determine this as a historical fact, we fail to reach the history, with which it is concerned.' (Heidegger, 1989, p. 80) The gods essence between absence and presence; they are having-been (*ge-wesen*). In order to respond to it, language must arise from a knowledge of the having-been, that is out of mood. For that reason, Heidegger says that 'fundamental mood … [opens up] the world which in the poetic saying receives the imprint of be-ing [*Seyn*]' (Heidegger, 1989, p. 80)

Fundamental mood opens up the having-been, which is responded to in art and in poetry, or rather the poetic response is one which involves a setting in work of the original having-been addressed by the having-been. The work is not a dissolution of the primordial language, but a rejoinder to the kairos, to the unique, a response which brings the kairological movement of the unique to a stand. The response can never completely correspond to the claim of the origin. Does that mean that truth as fulfilment acts here as a regulative idea? I cannot here engage in a discussion of Heidegger's critique of Husserl's concept of truth as fulfilment of an intention. All that can be pointed out here is that truth is not being understood as fulfilment, but as correspondence – *Entsprechung*. There is no time – understood chronologically – to fulfil the claim, because the response does not relate to a not yet or a no more, but rather to the origin.

The work refers to a having- been which it can never fully appropriate. In the work – and only in the work – the human being senses this having-been in its inappropriableness. As such, the position of the work is remarkably ambivalent. It forms a way to the having–been, and at the same time reveals the having-been in its absence. It stands to the having-been in an acausal relation, as not being able to appropriate it and yet being a response to the having-been. The relation of the work to the having-been is a *free* relation, a *free response*. Freedom claims the human; it calls for a response. As responding to freedom, the human being can understand its Dasein only in relation to the work, because the human being experiences itself there as founder and grounder. To ground something, the grounder must respond to that which cannot be given grounds, to the unfathomable, i.e. to the earth. But the earth is at the same time the within and upon which of human dwelling, and as such historical. Historical worlds have left their trace on the earth. The response to the earth is a response to the earth imprinted with world, to the historical earth. The work as response responds then to the earth imprinted by works. (cf. Heidegger, 2009, pp. 71–2; GA 38, pp. 84–5)

The danger of a regress is apparent here. But the work – or as Heidegger says 'great art' (Heidegger, 2002, p. 19; Hw, p. 25) – is unique. It resists all regression. The uniqueness of the work lies in its possibility, which is traced in its actuality. Only what is unique requires a response, because only this resists each attempt at appropriation. Yet, preservation is essential to the unique, indeed preservation begins with the work itself, in preserving the occurrence of the emergence of world in its strife with earth. This strife is the unconcealment of being, which has no ground, but rather an origin. The unconcealment of being is a transformation; the work arises from the striving to encounter this rupture through action, and the desire to preserve truth and form a continuity. Between the unconcealment being and the work is the space of transformation between possibility and actuality, motion and repose.

II.3 The futuricity of the work

The work is both a response to the having-been and a promise of a future. Heidegger speaks of a 'turning around' (*Umkehrung*) of temporality from having-been to the coming-to (future). (Heidegger, 1989, p. 103) This sudden change on which the work as the strife of world and earth is based, is itself poetic, because poetry is the bringing to language of the unique occurrence of this strife of truth, '[T]ruth is the illuminating clearing [*Lichtung*] and concealing of that which is, happens through being poeticized.' (Heidegger, 2002, p. 44; Hw, p. 58) In poetry, this sudden change happens in fundamental mood. The fundamental mood of poetry is sorrow. (cf. Heidegger, 1989, p. 81) Sorrow is the mood of metamorphosis. Metamorphosis makes necessary a closing off of the past, a taking leave of it. (cf. Gadamer, 1993, pp. 148–9) Such a taking of leave is 'grieving', and the poet feels this sorrow most acutely in foregoing the gods. Such foregoing is not done in hopelessness, but rather in awaiting the 'new beginning'. (cf. Heidegger, 1989, pp. 82, 93, 103–4) At once in sorrow, the having-been and the future is opened up, because the sorrow concerning the flight of the old gods and the sense of the godlessness of the present happens only in the preserving of the divinity, i.e. the possibility of the gods as a possibility of human action.[41] The gods are a possibility which remain unfulfilled. This lack of fulfilment is one which is not simply contingent, but is in principle such.[42] Hence, sorrow is the fundamental mood of poetry because it corresponds to radical possibility as a characteristic of all sudden change. In actuality the gods cannot return, because to do so would be for the gods to come into a proximity which is not appropriate for them. Nevertheless, in such sudden change the holy can form a possibility of action in such a way that the human being enters into another world. Yet in this possibility there lies a promise which offers more than it can fulfil. Sorrow as opening a future consists in living with this possibility and this lack of fulfilment.

II.3.a The promise of the work

Heidegger understands the poetic in art as projection. (cf. Heidegger, 2002, p. 45; Hw, p. 58) But the future which is opened up in projection remains alien to the work itself: it is not the future of dealings with the work in its actuality, but that of the world which is opened up in the work. This is what is hinted at in the formulation I have employed above: to promise more than it can redeem. This promise is making a claim as to the future and, in so doing, founding a continuity into the future. It is a certain saying, contained in the work. This saying is a 'projective saying' [*entwerfendes Sagen*] (Heidegger, 2002, p. 46; Hw, p. 60). In 'On the Essence of Grounds', Heidegger understands projection on the basis of grounding. In the Artwork Essay, he continues to understand projection poietically, but this account is enriched by considerations of language and poetry. The saying of poetry does not narrate anything and does not create any chronology. It does, however, open up a time insofar as it brings to language that which is not objectively present, and which yet is manifest in the poetic saying as the tidings it give of itself in it uniqueness. The character of projection of poetry is this bringing to language of that which is to its outer possibilities and limits. (cf. Heidegger, 1987, p. 155; EiM, p. 119) This presupposes a liberation of language, which Heidegger expressed as follows in the 'Letter concerning "Humanism"': 'The liberation of language from grammar into a more essential framework is reserved for thought and poetizing [*Dichten*]'. (Heidegger, 1998, p. 240; Wm, p. 312 [translation modified]) That which is no entity, the nothing, is being itself. This fundamental word of philosophy and poetry – being – corresponds to no thing or entity. (Heidegger, 1987, pp. 89–90; EiM, p, 67) At this level, what is thought is not; but rather the possibility of new experience or better of a new *order* of experience. It is precisely because the language of prose fails to grasp being, that being must 'experienced anew from the bottom up'. (Heidegger, 1987, p. 204; EiM, p. 155) What has failed here is the language of philosophy – the language of concepts and grammar. It is in the face of this failure that the thinker turns to poetry.

What is of concern here is a unique occurrence, as a new order of entities can only arise in an initiating historical situation. Only when an old order fails can there be the occasion for the arising of a new one. The new, the initiating, withdraws from the human power of disposing and ordering; it is not a matter of the will, but rather an occurrence of truth – albeit one for which human beings can and should prepare themselves. The connection between the initiating occurrence and language is one which is an event of language itself, as 'language brings entities as entities, for the first time, into the open.' (Heidegger, 2002, p. 46; Hw, p. 59 [translation modified]) Entities are entities only in their difference from being, and language opens up this difference for the first time.

But language is not only the openness of entities, but also a 'tiding', it gives tidings to those who hear it. Human beings act in the knowledge that, from their actions, certain tidings can be preserved. Language does not only open up entities, it also contains the possibility of preserving that opening. But this tendency of language to preserve can only be realized when language is set in work. Immediately on speaking of the new experience of being (quoted above), Heidegger goes on to stress that this experience has to be set in work. (cf. Heidegger, 1987, p. 204; EiM, p. 155)

The question arises as to whether the tiding of the openness of being can be a faithful one. In other words, is the occurrence of opening up, the event of the unconcealment of being, set in work faithfully and preserved as a tiding? Heidegger does pose this question in a lecture course from the summer semester 1934 concerning the relation of the occurrence in which history arises and the tidings given of that occurrence, although unfortunately he does not answer the question directly. (cf. Heidegger, 2009, pp. 86–93; GA 38, pp. 90–9) We can, however, piece together his probable response. In that same lecture course, he says that history can occur without there being tidings of it, and it is not always only the unimportant of which there are no tidings. (cf. Heidegger, 2009, p. 86; GA 38, p. 90) This is a paradoxical statement. Normally we think an occurrence – in particular, a historical occurrence – to be that of which tidings are given, that which those who witness those things are most likely to speak about. As such, Heidegger's statement is only intelligible if there can be that in occurrences which does not appear as that which can be told, as facts to be given in testimony.

The genuinely historical is to be found in epochal moments, which however could only ever be experienced from a fictive standpoint outside of history. This is so because we never experience the moment in its presence. Every narrative of historical events is either before or beyond the moment. "The genuine tiding of history gives tiding precisely in setting us before the concealed. The enigma of the moment of vision is the tiding of the overpowering and inevitable.' (Heidegger, 2009, p. 136; GA 38, p. 160 [translation modified]) The work also lies outside the epochal moment. Because the work is within the tension between chronology and the kairological rupture, it can never portray faithfully the sudden change itself. It *promises*, however, to do just that. To the extent to which it is an opening up of world the work promises to grant insight into the occurrence of this opening. The work sets up (*aufstellen*) a world; a world breaks open in the work. In the case of the Van Gogh painting of the peasant shoes, a world is opened up in the artwork. The thing as it appears in the work allows a world to be seen. But what is thus promised is that the worlding of that world be seen also. This is so because the work does not simply reproduce the thing – if that were all it did then the artwork itself would simply disappear – it rather allows the thing to be seen in the occurrence of its appearing, that is in the happening of the world in which it is. In this, the artwork tears the human being out of her familiar surroundings and gives her the prospect to be placed in the presence of the opening up of a world through the actuality of the work. The understanding of the artwork as 'timeless' arises out of this promise. However, the opening up of a world is precisely not 'actual', but rather has the character of possibility. The future is contained in this possibility, not as that which is set in the work, but as an excess in the actuality of the work, which indicates itself in it, but is never bodily there in it.

That which the work promises is the world which is opened up in it. This world, in contrast to the world which is disclosed in equipment, is not an already existing order, but rather a new world. In the lack, which we experience in the work, the new is indicated, and in that we find the historicity of the work. Because of this lack, there is in every artistic realization – that is, in every work – an excess. Without this excess there would be no history, as the future would be derivable in principle from the already realized. In such a case, the indefiniteness of the future and the possibility of transformation which is inherent in that indefiniteness would have no ontological significance, and would simply amount to symptoms of the limits of human knowledge.

If the work promises more than it can redeem, this more is contained in the kairos. The new world which emerges in the kairos claims us, but also makes a promise. The work is promising – we may expect a lot from it. The world which is set up in it opens up not only the possibilities of dealing with the work, but also possibilities which go beyond the actualizing of the work. That which can be perceived in the work is the future of Dasein, of those to whom the work appeals, and who in that appeal find themselves in a new world in relation to themselves and to others. 'The world is the self-opening openness of the broad paths of simple and essential decisions in the destiny of a historical people.' (Heidegger, 2002, p. 26; Hw, p. 34) Such decisions concern the practice of the human beings and the possibilities of relating with others opened up in the work. The decisions which are made in the world set up by the work concern the practice of those in a finite world.

II.3.b Decision and knowledge

The question of knowledge, knowledge of the future, is crucial to any understanding of human action in relation to that future. Heidegger gives some hints as to how to approach that question at the end of the Artwork Essay, where he states:

> Such reflections [whether art is an origin in the historical existence of a people] cannot compel art and its coming-to-be. But this reflective knowledge is preliminary and therefore indispensible preparation for the coming-to-be of art. Only such knowledge prepares for art, the space, for creators, the path, and for preservers, the location. (Heidegger, 2002, p. 49; Hw, p. 64)

In a similar place in the first draft of the essay, he states, 'knowledge concerning essence is knowledge as decision.' (Heidegger, 1989a, p. 22)

The promise of the work and the worldly decision characterizes the crossing over of kairos and chronos, poiesis and praxis in the work. If the promise of the work is in principle not redeemable, this means that it opens a futural horizon which can never be fully closed, because a time will never come in which it can be definitively said that what was promised is now done. This is the poetic in the work, poetry in the essence of art. However, the knowledge of such possibility in art is not a promise, but a decision. This decision is not a decision within art, not one which concerns art as such, but rather a decision concerning the possibility of art in the future. With this question, Hegel is directly addressed. In the Afterword to the Artwork essay, Heidegger reveals the relevance of Hegel's thesis that art is, in terms of its highest vocation, a thing of the past. Heidegger makes clear that any genuine encounter with Hegel on this issue concerns the future, 'the decision about this judgment will be made, when and if it is made, from and about this truth of being.' (Heidegger, 2002, p. 51; Hw, p. 65 [translation modified]) This decision is not one which is made systematically at the end of the first beginning, but rather one which is made out of the other beginning, the new beginning, which Heidegger is pointing towards. (cf. Heidegger, 1999, pp. 218–19; GA 65, pp. 310–11) In this context, the decision about the future cannot assume a chronological continuity, but must rather practice an attentiveness of that which is hidden in all chronologies. Such attentiveness is a decision as it does not allow the future to be an unquestioned continuity of the past, but opens up the possibility of a future which arises precisely in contradiction to the past. This decision has nothing to do with a choice between various 'projects', as is assumed by those – both attacking and defending him – that interpret Heidegger decisionistically,[43] but rather is a choice concerning the mode of being temporal. Heidegger addresses this directly in the first Hölderlin lecture course, where he describes the temporal decision concerning time (*Zeitentscheidung*): 'This is what is to be decided: whether we decide for the authentic time of poetry with its having-beenness, future and present or whether we remain hanging on the everyday experience of time, which takes everything as only "historical-chronological"'. (Heidegger, 1989, p. 112)

Heidegger, as we have seen, translates *τέχνη* with 'knowledge' [*Wissen*], and such knowledge is 'the ability-to-set-in-work the being of any particular entity.' (Heidegger, 1987, p. 159; EiM, p. 122 [translation modified]) As such, knowledge is grounding, but at the same time discloses the absence of ground – the abyss [*Abgrund*]. Such knowledge is not of an object, but rather encounters the boundaries of knowledge, namely contingency. This contingency appears in the possibility and futuricity of Dasein. On the basis of this contingency, every world historical situation is unique in its own manner of being. In the uniqueness of such situations all presumptions are empty and all planning remains futile.

As Heidegger puts it, 'we cannot predict or plan with regard to the uniqueness of our world-historical situation'. (Heidegger, 1989, p. 184) Human Dasein as knowing with respect to kairological futures is a being in relation to becoming, a becoming which is not purely poetic, purely thoughtful or purely active [*tathaft*] becoming. (cf. Heidegger, 1989, p. 184) Such knowledge is as confined neither to the realm of poiesis, nor to the realm of praxis, but concerns the becoming of human Dasein.

The future of such becoming cannot be represented. It is uncertain and contingent. The futuricity of Dasein in such a kairological situation is uncertain, and as such its whole existence is in play. The whole existence is one which Heidegger understands as concerning the poetic, thinking and acting powers – *dichtende, denkende und handelnde Mächte* – manifest in historical becoming. What is needed for knowledge of such becoming is 'to experience the enigma of their [the poetic, thinking and acting powers] original belonging together and to form them into a new and until now unheard of structure of be-ing'. (Heidegger, 1989, pp. 184–5) What is at issue here is a future order of entities. Such a new order comes into question only in a kairos, which is also a crisis. A crisis concerns the whole; what was receives a new meaning in the crisis and becomes *essentially* other. (cf. Heidegger, 1999, pp. 208–9; GA 65, p. 295) A crisis takes from those affected the possibility to persist with the old. A crisis is a dividing, or rather a plurality, of divisions: between those who pose the essential questions, and those who simply parrot those questions; of those who pose the essential questions, they too are divided between those who pose them poetically, in thinking or in action, and who in corresponding ways respond to the essential questions.

The fundamental division is between the possibilities of rejoinders to the happening of being. The human being does not know which answer is 'correct'. Indeed, this manner of posing the problem is already mistaken because what is at issue is not to find the correct direction for the answer, but rather to be prepared responsively for the opening up of a new world. In this sense, we need to hear the etymological meaning of *Ent-scheidung* ('decision'). '*Ent-*' as a prefix means both to accentuate and to negate. Hence, in understanding knowledge as decision, Heidegger is accounting of it as an opening up of divisions, and hence the making possible an authentic response to the situation, one which can overcome those divisions. Knowledge, so understood, lets a new world arise out of the new 'structure' of being, as it comes to be in human poetizing, thinking and acting. Such knowledge is at the same time one which arises in sorrow as a knowledge of the necessity to take leave of the past. Such knowledge is in relation to the character of possibility of the future, the necessity of decision and the actuality of the work in which unconcealment is brought to a stand.

The knowledge of the future is thus not a knowledge about *what* will be, but rather a knowledge *that* the situation is one which calls for decision. Such knowledge is itself a decision; such a situation is only for those who are resolved to respond to it. This is already a decision for the time of kairos and a liberating from chronos. At the beginning of this section I cited without further commentary parallel sentences from the first and third (final) draft of the Artwork Essay, in which Heidegger spoke in the later version of 'preparedness' [*Bereitschaft*], and in the earlier one of 'decision'. In both cases, what is of concern is the preparedness for the karios, the sudden occurrence. The decision is the response to the situation, and such a decision can only be a response because it is not within human power to bring the situation about. We cannot know concerning this occurrence *when* it will happen. To be able to decide, though, we must be prepared and must persist in the decisive situation. This preparedness is not a knowledge of how to deal with the work, it is not a knowledge of making or of preserving, it is rather a knowledge which stands prepared to make or to preserve. This preparedness is openness for a time which is heralded in the work and which can be traced in it.

To recognize the decisive situation it is necessary for the human being to decide concerning himself, to find himself on the crossroads of the kairos. (cf. Heidegger, 1987, p. 110; EiM, p. 84) This is most radically so in the case of a revolution, and it is no accident that Heidegger speaks of revolution in the context of this discussion of deciding concerning oneself. In his lecture course from the summer semester 1934, which has already been mentioned, Heidegger speaks of this clearly in the context of the National Socialist 'Revolution', which will be discussed in the next chapter. (cf. Heidegger, 2009, pp. 44–5; GA 38, pp. 48) In a revolution, the self finds itself in question, and is placed by others in a challenging situation in which it must decide whether it wishes itself or not. As Heidegger puts it (notoriously) in the Rectoral Address, '[W]e will ourselves' (Heidegger, 1985, p. 480; R, p. 19) Once a revolution begins, those affected by it are drawn into it, and the question each one asks themselves, 'who am I', takes on a necessarily political significance. In a revolution, there is no longer a stable order which the self can rely on to secure its own 'identity'. The self cannot appeal to its place in such an order to justify itself, and is radically subject to others. Hence, characteristic of revolutions is the phenomenon of denunciations. We can see many examples of this, but that of the Jacobin terror during the French Revolutions remains the classic instance, where in the absence of an accepted order of behaviour and of self-identity nobody was above suspicion, and indeed the very claim to order itself became an object of suspicion. (cf. Furet, 1978, pp. 70–87)

The decision of the self concerning its own authentic self in the absence of established order, precisely in the kairological situation, occurs in relation to another. What this suggests is the necessarily ethical component of the futuricity, something which Joanne Hodge has shown very well. (cf. Hodge, 1995) The future of the work is excessive, and as such undermines the binding power of every plan, every relationship and every promise. On the other hand, such commitment over time, such pledging to continuity of past and future, is essential to engagements with others, as without it one cannot bear and responsibility for the other. This leads to an apparent aporia: on the one hand, futuricity opens up the possibility of ethical engagement with the other; on the other hand, it seems to undermine its very possibility, by making all engagements with others, and indeed even with the self's own future, impossible. The only way of avoiding this aporia, it seems to me, is if the future can be understood in terms of a crossing over of kairos and chronos, of praxis and poiesis.

The movement of the kairos is brought to a stand in the work and thereby the possibility of its preservation is opened up. But preservation of the kairos allows it to be absorbed back into the chronological. The kairos never totally disappears in the chronos, however; it is still traced in the work. To follow this trace is the task of thinking, which prepares for the unique. In this lies the possibility of the future. However, thought cannot abandon the dimension of chronology precisely because of its responsibility to justify itself before an other. Whoever understands the kairos in purely poietically terms fails to understand revolution. The task of thinking is precisely to bring to light the interplay of chronos and kairos, praxis and poiesis. The difficulties inherent in such a task lie at the basis of the problem of politics in Heidegger's thought, and perhaps are at the roots of the crisis in his thinking which became known as the 'turning' (*Kehre*).

5

The Time of the Work II: Thinking and Politics

Heidegger's thinking does not suddenly become concerned with revolution in 1933. His thought is from the beginning – that is from 1919 – a thinking of revolution. To think revolution is to think the possibility of radical, abrupt, sudden and transforming change, or rather to think such change as possibility. For Heidegger, the question of being articulates the basic insight that thinking is thinking in relation to a time and that the present time – Heidegger's present time – was one in which the positive possibilities of thought had being exhausted and a new thinking was necessary. To think such a situation is to think without content. Heidegger's thinking is peculiarly contentless, in the sense that there is nothing positive to take from the tradition, its concepts and structures of thinking need to be overcome, but anything with which they are to be replaced is not yet there, but rather yet to come. This is not a nihilism which Heidegger produces, it is not a nihilistic thesis, but rather thinking without the possibility of having any thesis with content. Thus, for all Heidegger's polemic against novelty, as indeed indicated precisely by that polemic, is his underlying directedness at novelty – what he comes to call the 'new beginning'.

As such, the criticism which begins with Löwith and which can be traced through much of the recent polemics against Heidegger, namely that he sets forth a nihilism which makes null any criteria for judgement, and that this then leads him to have no means of judgement against the Nazi regimes, is both correct and trivial: correct in the sense that Heidegger's claim is precisely that, in the exhaustion of the tradition, there are no criteria left to which a genuine and honest thinking can avail us of, but trivial in the sense that this merely expresses the crisis of our historical situation, as Heidegger understands it. To simply point this out is not to make one step towards any engagement with Heidegger's diagnosis of the historical situation in which we find ourselves. Furthermore, while such critiques rightly point to Heidegger's anti-modernism with respect, say, to subjectivism, liberalism and modern technology, it ignores the fact that Heidegger embraces that most modern disposition to the world, namely the revolutionary. If Heidegger does not prescribe how we – and he

– should act, it is because he has taken the revolutionary spirit of modernity to its ultimate conclusion: within a revolutionary situation, the past can offer us no model for action, it can only offer us that which can be repeated in a creative rejoinder.

In his own terms – that is, in the terms of his philosophical project – Heidegger's political engagement was, as he said himself, a 'great stupidity' (*große Dummheit*) because he mistook the Nazi seizure of power – really Hitler as an historical 'event' – as a genuine revolution, that is, as an authentic repetition of the original question of being.[1] In his political pronouncements in 1933–4, and in his lecture courses at this time, Heidegger uses the term 'revolution' freely to refer to, or to point towards, the events of 1933. However, such a use of the term 'revolution' simply reinforces a constant motif in his thinking which I have been charting, namely that of transformation, of the sudden and free emergence of a new order of entities, of origins and abrupt beginnings. This motif is carried on later, but the term *Kehre* ('turning') is already in use, for example, in his 1928 lecture course *The Metaphysical Foundations of Logic*. Although I have not explicitly discussed the *Kehre* in this book, it is in the background of these considerations. We can see this once we ask about the place of the *Kehre* in Heidegger's thought: it cannot be chronologically situated in Heidegger's thought, because it expresses the revolutionary claim which Heidegger made for his thought and for thinking as such from his very first lecture course in Freiburg. As Reinhart Mauer already saw:

> "*Kehre*" means …revolution and in such a way that by this term is not meant merely a political revolution, but at the same time a transformation of the position of human beings in relation to themselves and to the world, hence of the fundamental disposition which is conditioned in modernity by technology. (Maurer, 1970, p. 244)

In the Rectoral Address, Heidegger makes clear that this revolution involves a repetition (in his sense) of the Greek beginning. It is important though to realize – especially when the discussion concerns Heidegger's 'anti-modernism' – that 'revolution' in the sense in which Heidegger is using it is a modern and distinctly non-Greek idea.

This present chapter argues that at the core of Heidegger's political engagement is his understanding of the 'work' as political. The time of revolution is for him intimately associated with the time of the work. If we are to come to terms with Heidegger's political engagement, then we have to understand it in terms of the time of this work.

I Revolutionary time

In a lecture course from the winter semester 1937–8, Heidegger states the following: 'The *Futural* is the *beginning of all occurrence* [*das Zukünftige ist der Anfang alles Geschehens*]' (Heidegger, 1994, p. 35; GA 45, p. 36 [translation modified]) The futural – that which opens up as 'to come' – is precisely that which discloses the beginning. Heidegger goes on to account for this future in terms of the 'rank and extent of creative action', and then continues:

> because the beginning is the most concealed, because it remains inexhaustible and withdrawn, because, on the other hand, what has already been becomes immediately the habitual and because this conceals the beginning through its extension, therefore what has become habitual needs transformations – revolutions. The original and genuine relation to the beginning is therefore the revolutionary ...The beginning is therefore not preserved – because it is not reached – by the conservative. (Heidegger, 1994, p. 35; GA 45, p. 37 [translation modified])

Despite the fact that Heidegger is often classed as a 'radical conservative' (by Farias, Bourdieu, Wolin and Faye, among others), he disavows that title more than once. For him, the true revolutionary is not a conservative, is not seeking to return to the past, but rather seeks precisely to transform the past by repeating its origins, its beginnings, in a radically new way. Indeed, in this lecture course in particular, Heidegger is anxious to distinguish his own account of the relation to the beginning from the conservative relation to the past. The conservative, he says, 'holds tight only to that which has started in consequence of the beginning [*was zufolge des Anfanges begonnen hat*]', but to begin means 'to act and think out of the futural, the extraordinary'. (Heidegger, 1994, p. 38; GA 45, p. 41 [translation modified])

The thought of a coming revolution characterized Heidegger's thinking at least since 1919. He was not politically active during the 'revolutionary events' of 1918–20. However, his lecture course in the summer semester 1919 was concerned with the situation of the university, something which we know from his later activity as Rector was of general political significance for him. (cf. Heidegger, 1987a, pp. 173–82; GA 56/7, pp. 205–14) As we can see from the above quotations from almost 20 years later, Heidegger's understood revolution as a new beginning, which involved a rupture from that which the original beginning had set in play. The experience of time as that which allows access to that original beginning is a continual concern for Heidegger from the notion of 'life-intensification' in the 1919 course (Heidegger, 1987, p. 176; GA 56/7, p. 208) to the later account of the moment of vision. Such a view lies at the basis of his strategy of destructuring (*Destruktion*) and is a rejection of reform. Reform remains imprisoned in tradition. The destructuring repetition of history has essentially the character of revolution because it seeks a liberation from the habitual in order to reconnect with the originating freedom of the beginning. Such a liberation breaks not only with the linear view of history as chronological narrative, but also from any cyclical account of history. This is so because what arises from the repetition of this beginning is nothing that can be already found there. In thinking, a split arises which the thinker can only encounter without knowledge – that is without technē. The technē of philosophy breaks down because it seeks to respond to that which it does not know, namely that which originates thought, but which cannot be known by what it has originated, because the latter has covered it up.

It is important to note here that the concept of revolution which Heidegger employs and which is fundamental to his thought is practically the opposite meaning which the word had from Antiquity up to the French Revolution. Originally, the word 'revolution' was an astronomical concept meaning the cyclical movement of the planets. (cf. Arendt, 1973, pp. 21–2; and Koselleck, 1985, pp. 39–54) In political terms, this meant the restoration of the original and proper order. This, indeed, is the most natural meaning of the word, if we listen to it etymologically.[2] Implicit in both meanings of revolution is a certain history of decline, indeed decadence: a revolution in both senses is motivated by the decadence of the existing order. However, the diagnosis of this decadence differs: the original meaning of revolution is premised on the assumption of an original, proper order which has been corrupted, usually through the actions of malicious people or institutions. The modern meaning of revolution assumes that the old order is exhausted, that far from pointing to the need to return to its original state, the decadence of that order indicates its inner bankruptcy and that the true response to that decadence is to build a fresh order on new and genuine foundations. Heidegger identifies the 'conservative' with the first meaning of revolution: he is the one who, in the face of present decadence, seeks to hold fast to a past and presumed better order. In contrast, he denies that any such return to a past actuality is either possible or desirable, and seeks through a destructuring repetition of the having-been to respond to the origin, which is not past but future in the sense of the possibility to be, which arises in and through a revolution. The coming new order is in this sense not producible in the classical sense of technē because there are no possibilities from the past which it can go back to as models for imitation (*mimesis*).

The change in meaning of revolution mirrors a change in the meaning of action. For Plato, the 'revolution' of the stars is the image of eternity (*αιών*). (Plato, *Timaios*, 37c6–38c3) Time has being only as an image of eternity. The order of the heavens (*κόσμος*) allows for nothing new to arise, 'nothing', that is, which is not an image of that which already is. No political order can last for longer than a limited duration because it is a transitory, passing phase of human life. (cf. Plato, *Republic*, 545c8–569c6) The transition from one phase to the next does not mark a rupture in the natural cycle, but rather *confirms* it. (cf. Meier, 1984, p. 669) The possibility of kairos is not there because every apparent rupture is already prefigured in a 'original past' order – the a priori perfect. In the light of this, it is understandable how the polis can be seen as a work, and knowledge concerning that work be interpreted as technē.

If now, however, the concept of revolution is transformed, the question arises whether the Platonic understanding of kosmos and time must shift as well. Is it possible to understand time platonically and at the same time to think the possibility of the new in the sense of the revolutionary new? What is striking about Heidegger's political engagement is the implicit and explicit appeal to Plato. The Rectoral Address with its division of service into labour, armed, and knowledge, as well as the platonic understanding of the relation of philosophy and politics with the philosopher leading the politician, even the final quotation from Plato (if a violent one) in that address (Heidegger, 1985a, pp. 476–7, 480; R, pp. 15–16, 19): all this displays a fundamental evoking of Plato in the face of what Heidegger saw as a decisive time, a kairological moment. Yet, this appeal to Plato is in effect a failure of nerve: a failure to face up to what the confrontation with the new entails.

II The work of the polis, the work of thinking

The work is, in Heidegger's understanding, unique. Such uniqueness is characteristic also of a people (*Volk*). A people cannot be defined as an objectively present thing: it is not a what, but a 'who'. (cf. Heidegger, 2009, p. 60; GA 38, pp. 68–9). A people are those who understand themselves as 'we' in terms of a common historical moment. Being a people is a decision, a decision with respect to a historical situation in which, as a people, 'we' find ourselves. As such, the uniqueness and singularity of a people has nothing to do with race.

There is no actuality to point to – neither race, nor citizenship, nor cultural traditions, nor activities etc. (cf. Heidegger, 2009, p. 57; GA 38, pp. 61–2) – but rather a decision concerning a 'mission and vocation' (*Auftrag und Sendung*) in which we find ourselves. (cf. Heidegger, 2009, p. 112; GA 38, pp. 127–8)

It is unfortunate that in many cases those who criticize Heidegger's Nazi engagement do not exercise the principle of charity in their interpretations and seem too easily to move from disinterested philosophers to prosecutors. A case in point is with respect to the concept of 'people'. It seems clear that Heidegger cannot be accused of biologism with respect to race or people, and as such appears to part company with the Nazis on a fundamental issue. In response to this, Tom Rockmore speaks of Heidegger's 'metaphysical theory of racism' (Rockmore, 1992, p. 59).[3] But, as Julian Young points out (Young, 1997, pp. 36, 44), racism is irreducibly a biologistic concept and as such to speak of metaphysical racism is, strictly, to speak nonsense. Furthermore, even if we distil the meaning from the phrase, namely 'an exaltation of the Germans specifically in virtue of their belonging to the German people' (Rockmore, 1992, p. 59), this misses the fundamental point that it is not the actual belonging to the German people which is at issue for Heidegger, but the decision to join together in a historical mission. That Heidegger displays a certain chauvinism with respect to the German nation and language is clear and that he, sometimes uncritically, takes up the Romantic myth of a special bond between the Germans and the Greeks is equally so. It is important, however, to see that this is a 'spiritual' bond; what Heidegger sees in the German language and history is the possibility to take up a historical mission. That possibility, however, is not in the interests of Germany in relation to any imperialist ambitions, but rather in the interests of the 'west' (*Abendland*), which for Heidegger is in the interests of humanity as a whole. (Heidegger, 1985a, p. 483; R, p. 23)

There is nothing either racist or fascist about this.[4] What is important is that the singularity of a people has to do with its particular historical mission and that this mission, in the case of the Germans, has to do not only with politics, but also with the history of being itself. The political decision of the Germans is one which has to be understood ontologically because it concerns the response to the unconcealment of being, the response to truth in its fundamental sense. The singularity of a people is at the same time the space of the being-with. For the Greeks, this space was called the *πόλις*. Heidegger understands the latter as the 'place [*Stätte*], the there [*Da*] wherein and as which Da-sein is as a historical.' (Heidegger, 1987, p. 152; EiM, p. 117 [translation modified]) A people is tied to such a place, which forms its home. This means that such places are not simply empty space in which the history of a people takes place, but rather, that which is moulded by a people, which in turn is moulded by it. Aristotle had already made note of the importance of place for the history of a people. According to him, territory is the hyletic basis of a polis. (Aristotle, *Politics*, c1326b27–1327a10; cf. Schwan, 1963) Aristotle conceives the polis as a work, and as such understands it on the basis of poiesis. As Arendt points out (Arendt, 1998, pp. 194–5), this reflects the political experience of the Greeks: for the Greeks, the making of laws was not a part of political action. The laws were understood in analogy to the city walls, as that which first made political action possible and as that which had to be made, and making is the activity of the craftsman, the *τεχνίτης*.

Law in this sense is closer to what we would understand as constitutional law, as that which founded or instituted the polis. Political praxis required poiesis to create the space for it. The metaphysical assumptions here are evident: to constitute the polis was to take material and form it according to a certain structure, the constituted means then something similar to form (*eidos*) or shape (*morphe*). (cf. Schwan, 1963, p. 90) The laws are the form of the polis; the people, the territory etc. is the material. In that case the polis is a work for Aristotle and as such praxis within it does not take its telos from itself (as praxis is otherwise understood to do), but rather receives it from that which has been set in work by the lawgivers. Analogous to the manner in which the polis has its visible boundaries in the wall, political praxis has its end (*τέλος*) in the laws, as Heraclitus suggests. (DK, B114) Thus the goals of political praxis do not lie in themselves, but rather in the poietical projection of a particular order, which must be seen as something constant in the polis, while praxis remains beset with inconstancy and transitoriness.

It is a platonic motif in Aristotle that he sees the polis as a work. In this he partakes of a dream of Western philosophy to conceive of the city state as an artwork. (cf. Lacoue-Labarthe, 1990, pp. 105–6) The question which must concern us then is whether Heidegger too fell victim to this 'fiction of the political' as Lacoue-Labarthe terms it, or whether he nonetheless glimpsed the possibility of action beyond the work and poiesis which would be faithful to that unknowing characteristic of the revolutionary situation. At this point it is necessary to recall the possibility of a non-technical theory which we discussed in Chapter 2. At the same time, the considerations of Chapter 4 on the concept of work and the problem of grounds need to be examined. On the basis of the latter, it follows the non-technical theory ought not be merely a praxical 'theory', but must involve a reinterpretation of both technē and phronesis. To pursue this theme further, it is necessary to discuss the relation of philosophy to poetry, as the latter is for Heidegger the essence of art.

Heidegger speaks not only of artworks, but also of *thought*works and *state*-works. It appears as if every relation of human beings to their goals is contained in a work. One may even gain the impression that any human mode of relation which cannot be set in work, such as interpersonal relations for example, is consequently devalued. (cf. Schwan, 1963, pp. 82–8) To examine this question thoroughly we must again investigate the relation of work and truth.

In the *Introduction to Metaphysics*, Heidegger states, 'unconcealment occurs only to the extent to which it is made actual [*erwirkt*] through the work'. (Heidegger, 1987, p. 191; EiM, p. 146 [translation modified]) In the almost contemporaneous Artwork essay he states: '*One* of these ways inwhich truth happens is the work being of the work'. (Heidegger, 2002, p. 32; Hw, p. 41 [my emphasis]) If we juxtapose these two sentences the question which emerges is whether and in what sense there can be truth which is not set in work.[5] This is to say that there may be an essentially non-poietic moment in truth. Truth happens only in and as the strife between concealment and disclosure; without such disruption being is silent. There is truth only when there are human beings. and the human being who experiences truth does so in the modes of creator or preserver. Through creating there is a work, or better, creating yields the work to the extent to which it is previously conditioned by the work. As Heidegger says, 'the essence of creation is determined by the essence of the work'. (Heidegger, 1987, p. 34; Hw, p. 46) In the work there occurs a strife between world and earth: 'Truth is present only as the strife between clearing and concealing in the opposition between world and earth. As this strife of world and earth, truth wills its establishment in the work.' (Heidegger, 2002, p. 37; Hw, p. 49) Can this strife take place without the work? Heidegger leaves open this possibility. In a marginal comment to a passage earlier in the Artwork Essay in which he says that the world and work belong in the unity of the work, Heidegger asks, '[O]nly here? Or here rather only in the mode of construction [*gebauten Weise*]' (Heidegger, 2002, p. 26) This, admittedly later remark of Heidegger's, nonetheless agrees with other hints in the text that the work is an excellent, but not the only, access to truth. The reference to 'construction' is in the context of the difference between praxis and poiesis of particular significance. The question is then, how to think this non-constructive access to truth.

The work does not communicate the true tidings concerning this or that, but rather 'allows unconcealment with regard to entities as a whole to happen.' (Heidegger, 2002, p. 32; Hw, p. 42) That is to say that through the work human beings are brought before the occurrence – or the event (*Ereignis*) as Heidegger remarks in a marginal note beside the word 'occurrence' (Heidegger, 2002, p. 32) – of truth as such. This occurrence is a rupture with normality, and through it self-evident truths are called into question. There are no truths to guide human beings anymore; rather human beings must let themselves dwell in the truth as an occurrence. Truth here cannot mean any actual truths, but rather truth as possibility, that is, as a possibility which the unique occurrence of the strife of world and earth opens up. This uniqueness of truth demands a decision, but decision understood as response. This decision does not concern a work in its actuality, but rather as possibility. The strife of world and earth places the human being before a decision: 'The dawning world brings to the fore that which is still undecided and without measure and decisiveness and thus opens up the hidden necessity of measure and decisiveness.' (Heidegger, 2002, p. 38; Hw, p. 49) The decision concerns the essence of a people and that means its future, the future order of its polis. In this sense the decision is political. But it does not follow from this that it is the politician who paves the way for such a decision. On the contrary, it is the thinker who does this. The thinker bears the care for art and politics.

The questions of the thinker are philosophical and as such necessarily concern (if only indirectly) the task of philosophy itself. Heidegger repeatedly dismissed the sectioning off of philosophy into ethics, ontology, political philosophy, aesthetics etc. as evidence of its decline and decadence. The questions which Heidegger poses of art concern, rather, the question of being.[6] But what this means then is that the question concerning art concerns the historical people for which the artwork is, because it concerns a historical decision as to how to respond to the unconcealment of being disclosed with the strife of world and earth in the work. As such, the thinker is concerned not with the work as such, not with making as such, but rather with the way of being of a people in the world which arises in the strife with earth in the setting-in-work. His task is not to create anything, but to know the occurrence of truth and to prepare a decision concerning it, namely how this occurrence can and should be encountered. The thinker does not ground anything. Heidegger, however, speaks of the truth of thinking as set in work in the 'work of the word' (Heidegger, 1987, p. 191; EiM, p. 146) Although it is undoubtedly the case that there are philosophical works, the question which is of concern to us is whether this is essential to philosophy as action. In the 'Origin of the Artwork', Heidegger describes the way in which the truth of thinking occurs in the following manner: 'a still further way in which truth comes to be is in the thinker's questioning, which, as the thinking of being, names being in its question-worthiness [*Frag-würdigkeit*]' (Heidegger, 2002, p. 37; Hw, p. 48) The question-worthiness of being makes manifest in an actuality – in the actuality of the work; but the task of philosophy is to transcend the actuality of the work towards its possibility. Such a questioning explodes all normality, because the latter remains subject to concern for entities. Being is alien to this. Indeed, being is alien to the human being 'in that they turn away from being, because they do not grasp it but suppose that entities are only entities and nothing more.' (Heidegger, 1987, p. 130; EiM, p. 100) This 'nothing more' is that which interests the philosopher, as Heidegger makes clear in his lecture 'What is Metaphysics?' (Heidegger, 1998, p. 84; Wm, pp. 105–6) For this reason, Heidegger describes philosophizing as 'an extra-ordinary questioning regarding the extraordinary'. (Heidegger, 1987, p, 13; EiM, p. 10 [translation modified]) Such questioning transcends the ordinary due to its prior fallenness. It is directed not so much at the work as at that which is manifest and traced in the work: the setting-in-work of truth, truth not as order but, as a bringing into order. This bringing into order Heidegger names the 'destiny of being' [*Geschick des Seins*].[7] In this sense the philosopher questions regarding origins. Thus, she asks not about the past, but about the having-been; she questions in the mode of repetition. As we have seen, genuine repetition is transformative and this is true most particularly with regard to the question concerning being: 'To question: how is it with being? – that means nothing less than to re-peat the beginning of our historical-spiritual Dasein, in order to transform it into an other beginning.' (Heidegger, 1987, p. 39; EiM, p. 29 [translation modified]) In this sense, philosophy is an action which questions beyond the work in preparing for an 'other beginning'. It is the 'unknowing knowledge' concerning the kairos of the beginning, which must remain always alien to the work. Strictly speaking, philosophy neither creates nor preserves, but rather waits for the occurrence of being and prepares for it. As Heidegger puts it, somewhat wistfully, at the close of the lecture course, *Introduction to Metaphysics*: 'Being able to question means being able to wait, even a whole lifetime … But the essential is not number; the essential is the right time, i.e., the right moment, and the right perseverance.' (Heidegger, 1987, p. 206; EiM, p. 157)

But questioning concerning being is not purely a philosophical affair. Heidegger speaks in that same course, of the 'fundamental poetizing-thinking experience [*dicthtend-denkenden Grunderfahrung*] of being' (Heidegger, 1987, p. 14; EiM, p. 11[translation modified]) Our concern here is the mode of action of philosophy and of poetry, which in the end is a question concerning temporality. Again here we have to face Plato, in particular Book 10 of the *Republic*.[8]

The question here of the relation of artist and poet is political, and the mode in which it is posed assumes as certain account of time: the poet remains in the realm of the temporal, while the philosopher is inspired in such a manner that he can glimpse the eternal. (cf. Plato, *The Republic,* 484b5, 500c3) When this temporal distinction is called into question – or if it is relativized at all – the danger is there that the difference between poetry and philosophy would disappear. This danger threatens Plato's own texts which are themselves poetic and this is consistent, because the philosopher can only strive for knowledge of the ideas and, in this striving, his role in the city – in the polis – remains uncertain and indefinite. (cf. Rosen, 1988, pp. 5, 25–6) Poetry is a political competitor for the philosopher because the latter uses and is forced to use poetic language.[9] Hegel set himself the task to overcome this danger and to found a philosophical system in which philosophy can secure itself against the poetic. In the *Phenomenology of Spirit*, he claims to have overcome philosophy as a mere 'love of wisdom' and to have reconstituted it as a system of actual knowledge. (Hegel, 1977, p. 3) In this attempt, the temporal (the historical) is sublated into the eternal whereby the real is understood as the sensible appearance of the idea.[10] This conception is not a break with the platonic thought as technē but rather precisely its confirmation. This is so because Hegel's attempt is to sublimate the discontinuity of revolution and the failure of knowledge, understood as technē. Heidegger, on the other hand, is concerned with thinking discontinuity itself.

What follows from this is that, for Heidegger, against Plato and Hegel, philosophy and poetry do not differ in terms of their temporality. The temporal decision which Heidegger speaks of in his first Hölderlin lecture course is not a decision against poetry; on the contrary, poetry projects the other beginning. Hölderlin as the poet of poets thinks beginnings originally. (cf. 1989, p. 269) Philosophy and poetry are both modes of acting making rejoinders to the unconcealment of being. The difference between them has to be in their ways of rejoinder. Heidegger states the difference in the following manner:

> The basic mood and that means the truth of the Dasein of a people is originally founded [*ursprünglich gestiftet*] through the poet. The Be-ing of entities so disclosed is conceptually grasped and structured [*begriffen und gefügt*] and in that way opened up through the thinker. (Heidegger, 1989, p. 144)

Thinker and poet depend on one another. The poet founds the truth of a people and also a historical world, as the truth of a people is its history. With resonances of what we found already in Aristotle, Heidegger speaks of poetry as the founding of be-ing as the originary 'legislation' [*Gesetzgebung*] (Heidegger, 1989, p. 258) Poetry is not in the situation to perform this task on its own. The new historical world 'can only come to be when poetry comes to the power of its essence and this poetry develops itself through the rigour and clarity [*Härte und Bestimmtheit*] of thinking-questioning knowledge.' (Heidegger, 1989, p. 221) So without philosophy, without the rigour and clarity of thinking-questioning knowledge the power of poetry does not come into play. At the same time, philosophy is peculiarly passive with respect to poetry. Only in the experience of founding and grounding does the human being come before the abyss, the occurrence of being, and this experience is gained only in the coming of poetry to language. Philosophy, as Hegel also saw, is therefore in a literal sense re-flection or even more clearly in German *nach-denken*, thinking which comes after, which comes subsequently. Heidegger, in reference to Hölderlin, characterizes such reflection as remembrance [*Andenken*]. Poetry, as much as philosophy. engages in remembrance. The difference lies in the fact that poetry is through its own project *forced into* remembrance; whereas philosophy *begins* with remembrance. Poetry founds and is always forced back into remembrance of that foundation; philosophy thinks only beginning from that foundation, thinking it back into its own origins. In this we can find an echo of the ancient idea – rehearsed by Plato – that the poet is inspired by muses whom he neither knows nor understands, and needs the philosopher to grasp and clarify his words.

Heidegger thinks poetry as foundation (*Stiftung*) and as such he thinks it on the basis of poiesis. He understands poetry as grounding and philosophy as thinking the ground of grounds. In comparing the relations of Hegel and Hölderlin to Heraclitus, Heidegger says, 'Hegel glimpses backwards and closes off [*abschließt*], Hölderlin looks forwards and opens up [a world] [*aufschließt*].' (Heidegger, 1987, p. 126; EiM, p. 96 [translation modified]). Although Heidegger always seeks to maintain his distance to Hegel, it would appear that reflection, i.e. subsequent thinking – *Nach-denken* – of a Hegelian type remains paradigmatic for philosophy and thinking and precisely distinguishes the latter from poetry, for which Hölderlin is paradigmatic. Poetic thinking is a technē, and technē is the 'ability-to-set-in-work' [*Ins-Werk-setzen-Können*]. (Heidegger, 1987, p. 159; EiM, p. 122) Poetry is knowledge and action in the sense of poiesis. A world is opened up in the work, but without philosophy the being-possible of this world remains opaque. Philosophy thinks behind the work to the occurrence of truth, but also past the work to the possibilities of the world which are accessible through it and which are traced in the work. Philosophy sets free the possibility of transformation, which is brought to a stand in the work. In doing this, it raises the transformation of entities as a whole, which is presaged in poetry, into the dimension of decision. Philosophy limits itself to prepare for the decision, so understood. In this way, the task of thinking is always marked by subsequentiality. The philosopher waits on and for the founding, on and for the grounding and the work of the poet, in which access to the truth of a people is opened up. The philosopher's thinking is a constant return to the source of the founded, to poetry.

III The creators

According to Taminiaux, the difference between praxis and poiesis disappears almost entirely from Heidegger's lectures in the middle of the 1930s. The thinking of being comes to be understood as the highest praxis and the highest poiesis. (Taminaux, 1991, p. 216) The tendency, which Taminiaux is remarking on, is certainly evident in Heidegger's writings and in fact helps make understandable his engagement with the Nazis. But Taminiaux's reading is one-sided and ignores the manner in which Heidegger is, in fact, rethinking the relation of praxis and poiesis in relation to the 'work'. There is a tension here in Heidegger's thinking which is brought out above all in the account he gives of the 'creators' (de *Schaffenen*).

> The original, historical time of peoples is ... the time of the poet, thinker and creator of states [*Staatsschöpfer*], i.e., those who genuinely ground and establish [*gründen und begründen*] the historical Dasein of a people. They are the genuine creators. ... The times of creators ... tower above the mere sequence of busy days. (Heidegger, 1989, pp. 51–2)

Kairological times give occasion to creation. Only the creators know how in such times it is possible to act and how we are obliged to act. To use one of Heidegger's favourite metaphors, they are on the summits above the plains of the everyday. The creators live in the vicinity of the abyss, as only in relation to grounding does the lack of grounds – the abyss, appear. As such, creation characterizes the abysmal essence of historical humanity.

It is in this context that Heidegger cites the choir of Sophocles' 'Antigone' referring to the human as *τὸ δεινότατον*, which he translates as 'the most uncanny of the uncanny' (Heidegger, 1987, p. 149; EiM, p. 114 [translation modified]) The *δεινόν* is the terrible in the sense of the 'overpowering power' [*überwältigenden Waltens*] and the overpowering are the entities as a whole. It means also the 'violent' [*Gewaltige*] in the sense of the one who uses violence. It is important to be clear here that while we translate '*Gewalt*' with violence, the German word has a much wider meaning, with an emphasis not on violation, but on the opposite, the legitimate exercise of power. That Heidegger wants us to hear this in *Gewalt* is clear from the word play with '*überwältigenden Waltens*'. The exercise of power by the human being makes him terrible, because it places him in relation to creation, something which is almost godlike – it places him, in other words, in the situation of the demi-gods. In this sense, the human is 'actively violent [*gewalt-tätig*] insofar as the use of power is the basic trait not only of his action but also of his Dasein.' (Heidegger, 1987, pp. 149–50; EiM, p. 115 [translation modified]) Heidegger makes clear here that 'violence' (*Gewalt)* has nothing to do with 'arbitrariness' [*Willkür*] or 'brutality' [*Roheit*]. Rather, violence is essential to the uncanniness of the human. The human being can overstep the boundaries of the normal and the familiar. This means that as uncanny the human is without state – *ἄπολις*. (Heidegger, 1987, p. 152; EiM, p. 116) Heidegger hesitates in using 'state' as a translation for *πόλις* is rather the 'place, the there, in which and as which Da-sein is historical.' (Heidegger, 1987, p. 152; EiM, p. 117–18) As the most uncanny, the human being is a creator because he is without city and place and must first create these.

The consequence is apparently clear; the model of action for the thinker, poet and statesman is a poietical one: kairological time is, it would seem, exclusively the time of making. Heidegger is understanding action here in terms of a revolutionary situation. For the most part, the human being lives in a *πόλις*. Only when the *πόλις* loses its stability and its order breaks down does the human being experience the contingency of this order and the fragility of chronology. Heidegger's interpretation of the 'Antigone' relates to the contemporary situation in Germany; in the summer of 1935, Heidegger still believed in National Socialism as the possibility of a German revolution, although one which had not as yet been realized. A semester before he had said, 'there is no reaction, because there is no upheaval [*Umwälzung*] (revolution), and there is none because no one has yet grasped where one should begin.' (Heidegger, 2009, p. 65; GA 38, p. 76 [translation modified]). In 1936, in conversation with his former student Karl Löwith during their final meeting in Rome, the latter reports him as saying that 'one had to only hold out long enough.' (Löwith, 1993, p. 142) And although the reference to the political situation is more oblique in his 1937–8 lecture course, the references there to revolution and the preparedness for the historical moment make unmistakable reference to that situation:

'historical reflection [*Besinnung*] works towards a preparedness for a historical existence which lives up to the *greatness* of fate and the peak moments of be-ing [Seyn].' (Heidegger, 1994, p. 50; GA 45, p. 55) The revolutionary situation is represented in such a way that within it there is no further possibility of praxis. In Greek terms, praxis assumes a grounding act; only a *πόλις* offers the opening, the space, where praxis-like relations are possible. The *πόλις*, however, must first be founded and a creative act is needed to found and establish it. Heidegger is indebted here to the Greek manner of thinking the *πόλις*, which we have already discussed. While Heidegger does reinterpret poiesis and in so doing allows its kairological qualities to becomes apparent, he remains indebted to an understanding of truth as technē and poiesis – the knowing and acting in relation to the setting-in-work. Revolutionary action in that context is not action in the mode of possibility, but rather action as actualization under the rule of technē, of grounding and establishing. The thinker in thinking the abysmal is compromised by the thinker as creator, who in cooperation with the poet and statesman is understood as engaged in the task of founding and establishing the polis.

Before exploring the implications of this, however, it is important to point to the ambiguities in Heidegger's account.

A revolution is a violent process. This violence is firstly apparent in the collapse of the old order. Such a situation brings human beings before the transformation of being. But in this situation the human being does not stand in the first instance as a creator, because every revolution is characterized by a peculiar lack of knowledge, a lack of knowledge as to how it is possible be engage with a kairological situation. This lack of knowledge places those in this situation beyond technē. There is for action no more models, these need first to be made. The normal rules and conventions for engaging with other people become no longer valid. Relations to others are no longer mediated through works, because the boundaries between producers and users, poets and politicians, thinkers and craftsmen are no longer in force. According to Heidegger:

> the gods, the temples, the priests, the festivals, the games, the poets, the thinkers, the ruler, the council of the elders, the assembly of the people, the army and the fleet. All this ...is political, i.e., at the site of history, provided there be, for example, poets *alone*, but then actually poets, thinkers *alone*, but then actually thinkers, priests *alone*, but then actually priests, rulers *alone*, but then actually rulers. (Heidegger, 1987, p. 152; EiM, p. 117 [emphasis in original)

But the corollary would be that outside the city the poet is never only a poet, the thinker never only a thinker, the ruler never only a ruler. What occurs in a revolution is not freedom but equality. All hierarchy, which lies at the basis of the state understood as a work, is destroyed.[11]

In such a state of equality, the knowledge which guides action can no longer rely on the rules and conventions of relations already laid down. In this context, action requires a constant openness and nimbleness in order to find new ways or to make them. In short, flexibility is required. Flexibility in this sense captures the sense of resoluteness in *Being and Time*. (cf. Sitter, 1970, pp. 530–2) As Heidegger states, 'the certainty of resolution means *keeping oneself free for* the possibility of *taking back*, a possibility that is always factically necessary' (BT, p. 284; H, pp. 307–8), i.e. it is required by the factically changing situation. If resoluteness opens up possibilities as well as closing them off, then it must remain always open for changes in the situation. This means factically that the resolute Dasein acts always in responsibility before the other; the change in the factical situation can only mean that the relation to the other undergoes a change within the situation in which Dasein finds itself. It is important to stress – something which is not always taken account of – that Dasein in resoluteness is not thrown back on its own individual self, but is immediately dependent on others; 'resoluteness ... pushes [Dasein] toward solicitous being-with with the others.' (BT, p. 274; H, p. 298). But in a revolutionary situation, those relations with others are without rules. In this situation – in this moment of decision (cf. Heidegger, 1994, pp. 51–3; GA 45, pp. 57–9) – Dasein is challenged to be towards the other as an other. The mediating actuality of rules, conventions and impersonal structures – works – has collapsed. In this situation there are only face to face relations where nothing outside them remains to mediate the 'limit situations' with others. Neither 'I' nor the other decide to act without rules or conventions, rather the decision is a response to that which happens between us. What happens here is a transformation of being which manifests itself as free-releasing possibility. This situation is not a 'zero hour' or indeed a 'ground zero' because the having-been essences still. But that having been is not recognizable or identifiable with what we are used to because the transformation of being is a metamorphosis through which even the assumption of mutual comprehensibility is not assured; even language itself, though the same, has been transformed.

The uncertainty of such a situation seeks new rules. It is not by accident that the American and French revolutions were intimately connected with the creating of new constitutions. This requires a setting-in-work, it requires knowledge as technē. But in that revolutionary setting-in-work there are no longer poets, thinkers or statesmen, as these very distinctions need to be produced again. For Heidegger, the poet, thinker and the statesman have a place at once inside and outside the polis.[12] This 'at once' or 'at the same time' must be understood kairologically. As creator, the thinker is a thinker only in the polis, but in that moment she ceases to be a creator. She has co-created an order in which she can live; previous to this creation she could not be as a thinker. That does not mean that there is no thinking outside the polis, only that outside the polis it plays a different role. This is so because in a revolutionary situation, thinking is unknowing. There is for thought no having-been order which can be imitated. As non-mimetic thinking is unknowing, it is accomplished outside of technē: it is a thinking of revolution (subjective and objective genitive). It is not thinking concerning a constitution to be created, but a thinking of the possibility of transformation, which responds to this possibility without knowing how or why an answer is possible or called for. It is a thinking with the sense for the new. Such thinking is, however, praxis not poiesis.

Against this Heidegger stresses that the thinker in a revolution is a creator, who thinks in relation to a work. Therein lies an ambiguity in his understanding of philosophy: philosophy as questioning and philosophy as thought-work (*Denkwerk*). This ambiguity is fundamentally platonic. I have already alluded to the Platonism of the Rectoral Address as a failure of nerve on Heidegger's part and I will pursue that in the next section. But it is important in this context to see that the difficulty here is one which goes to the core of the self-understanding of philosophy as poiesis. Above all here the question concerns the unknowing of philosophy, proclaimed since Socrates but always almost immediately covered over by claims to know, by expressions of philosophy as technē – indeed the highest techne as Aristotle affirms. (Aristotle, *Metaphysics*, 983a 6–8) But philosophy as questioning begins and ends in unknowing and indeed before the motion of the revolutionary occurrence, the abrupt moment of the transformation of being (*ἐξαίφνης*), thinking becomes silent. When the bearer of historicity is a people, then such historical unknowing, such silence, concerns a people or a community.

In his book, *The Inoperative Community*, Jean-Luc Nancy criticizes what he calls 'immanentism'. By this, he means the representation of a community as becoming one, that is a community as a work. Against this, Nancy argues that community opens up necessarily a dimension of the outside, the being-outside-oneself. (Nancy, 1991, pp. 7–9) In this sense, the experience of a common essence is one of finitude, because in all speaking and thinking the human being is outside-himself and as such shares something with the other. Such a sharing with is a response; it responds to an occurrence. Community reveals finitude and the I's essential relation to others: 'It is the presentation of finitude and the irredeemable excess that makes up finite being: its death but also its birth.' (Nancy, 1991, p. 15) Singularity is preserved for human beings through their birth and in their death, but precisely in this they are outside themselves and dependent on their community. In this sense, Nancy understands singularity and individuation as fundamentally distinct, 'the *singular* being, which is not the individual, is the finite being.' (Nancy, 1991, p. 27 [emphasis in original]) Here, Nancy shows that the moment of exceeding is constitutive for community. This he then uses critically to show how immanentist views – such as those of the National Socialists (Nancy, 1991, p. 12) – strive for a becoming one.

Heidegger's account of the work led him to a view of the state as work and as such drew him towards such an immanentism. Nevertheless, he both stressed the place of the poetic – and the thinking – work and yet, the work for him is *one* moment – if a vital one – in the movement of truth. The 'beyond' of the work is alien to the work. Human beings are with respect to that which is beyond the work in relation to the work, but in thinking the human being relates to the work in its possibility in the abyss of the groundless ground and in the possibilities which it opens up but can never fully close off.

IV Philosophy and politics

One of the ironies of the debate concerning Heidegger's relationship with National Socialism is that, while many of his critics seek to protect philosophy from Heidegger's politics, they repeat the charge against Socrates of corrupting the young. We see this first quite explicitly with Jasper's letter to the Denazification committee, repeated by Habermas and others, and made quite explicitly by Faye. (cf. Dilthey 1993, p. 149; Habermas, 1993, p. 190; Faye, 2009, p. 112) This should make us pause. Kisiel has already pointed to the Socratic parallels here. (cf. Kisiel, 1992, pp. 45–6) Philosophy since its inception has challenged the deepest held beliefs of society. Clearly Heidegger challenged our belief in democracy and – not unrelated to that – our faith in progressive rather that abrupt and disruptive change. Clearly there are fundamental dangers in such challenges as the experience of totalitarian regimes in the last century make abundantly clear. But to second guess thinking in the face of its risks is in the end to give up the philosophical enterprise itself. Hence, it is vital to consider Heidegger's relation to National Socialism in the spirit of Socrates and not in the spirit of the Athenian court.

Furthermore, the question has to be one of understanding Heidegger's politics on the basis of his philosophy, not the other way around. Thomas Sheehan proposed that, in reading Heidegger, his philosophical work from 1933 at least should be read politically. (Sheehan, 1988, p. 47) This seems an odd hermeneutical principle. It cannot simply be a principle with respect to Heidegger (unless one can show either that his philosophical works are not philosophical or that his political actions make him not a philosopher). As such, it would commit us, for example, to reading all Plato's texts in the light of his 7th Letter and his engagement with Dionysius II of Syracuse, or for that matter Aristotle's texts in the light of his defence of slavery, or those of Hegel in the light of his engagement with the Prussian regime. To some extent these strategies have been attempted, but these are hardly the most fruitful readings of these authors. In Heidegger's case, the publication of Faye's book in bringing such a political reading to its logical conclusion shows what this would mean: a reading of Heidegger which finally excluded him from the philosophical canon. (cf. Faye, 2009) The stakes here are indeed high: the political reading of philosophy taken to its logical conclusion undermines the place philosophy needs from which to speak of the political. Paradoxically, the judgement that the sometimes disastrous political results of philosophical engagement with the political is too great a price to pay for that philosophical space is itself a philosophical judgement, albeit a suicidal one.

What is striking about Heidegger's political engagement is its platonic tone. Taminiaux has already remarked on this and argued that Heidegger was guided by two fundamental Platonic ideas during his political engagement: the idea of the philosopher king and the understanding of the polis as a work. (Taminiaux, 1991, pp. 218–19) Heidegger in effect recasts these ideas in the light of his account of kairological temporality and his claim that the first beginning of philosophy was coming to an end in the crisis situation into which 'we' – we in the Western world – had entered. If we are to understand Heidegger's politics, his so-called Hitlerism, his actions and his application of his philosophical thought to the times, we need to be clear about this sense of crisis and Heidegger's 'platonic' response to it.

Heidegger's stress on the moment of vision is strong in his lecture courses from 1933–6. This is particularly the case in the summer semester 1934. To cite just one instance he states, '[I]n the moment of vision, in which we have grasped the We as a matter of decision [*als entscheidungshaftes*], the decision about our self-being has also been reached.' (Heidegger, 1994, p. 52; GA 45, p. 59) The moment of vision is a moment of decision in which we decide concerning ourselves. The nature of this 'we' is important here, because the crisis of which Heidegger speaks is a crisis for this 'we'. In fact, the crisis is not merely for this 'we', but is a crisis which concerns humanity as such. Although Heidegger only later develops a philosophically articulate critique of modernity and of modern technology, the crisis which he has been referencing since 1919 is one which concerns the historical destiny of the West. In point of fact, because Heidegger understands history in relation to philosophy, and human historicity as bound up with philosophy, his concern is with humanity to the extent to which that humanity is affected by philosophy. Faye understands Heidegger as making philosophy subject to the German nation, to in effect making the 'we' of philosophy equivalent to the 'we' of the German people, and as such undermining the philosophical goal of reason. According to Faye, this opposition to consciousness without roots in soil is anti-Semitic and a 'radically destructive challenge to an essence of man qua man.' (Faye, 2009, p. 36) Philosophy's fundamental question becomes one of the destiny of the German people such that 'the question of being has become, in Heidegger's teaching, beginning in 1933, explicitly a *völkish* question: it concerns exclusively the being of the German people and arises only with respect to this people.' (Faye, 2009, p. 92) Against a Cartesian cosmopolitanism then, Heidegger condemns any philosophy which is not based in the historical destiny of the German people. Faye quotes extensively from the 1934 course, which I have referenced in many instances above, to buttress his claim that Heidegger here betrays philosophy itself by making it nothing more than a means of furthering *völkisch*, and ultimately racist, ideas.

It is certainly the case, as Faye points out, that there is a definite move towards German philosophy and literature from the 1930's onwards: Heidegger lectures only on German and Greek thinkers and poets in his courses from this period, and furthermore the emphasis on the German people is hard to ignore. What Faye does not take sufficient note of, however, is that in the 1934 course Heidegger introduces 'we are the people' as a decision; it functions in this text not so much as a statement, but rather as a challenge. He is challenging his listeners to a decision concerning the 'mission and vocation' of the German people, a mission which he makes clear is to take its place at the centre of the West (*Abendland*). This mission is one which concerns the question of being not because that question is only one of concern for the German people, but because the Germans have the vocation – as the Greeks once had – to pose the question of being and to do so again in a new beginning. I submit that the latter claim should move us more genuinely to an ironical smile than to outrage; there is nothing in this claim which can in any way be said to be either racist or imperialist. That Heidegger attempts to retrieve the concept of race for his purposes as he undoubtedly does in this course (cf. Heidegger, 1994, p. 57; GA 38, p. 63) can be understood as naïve or indeed as opportunistic. I think it was both. But it is clear from this and from other texts of the period, especially the *Introduction to Metaphysics*, that Heidegger was employing these terms with respect to the kairological situation of the transformation of being and the revolution which in a political sense should respond to that transformation. Again the question of decision is important here: race is fundamentally a biologistic concept and we do not choose our biological givens. It is also in this context that we should read Heidegger's use of such terms as Liberal, America/Americanism, Russia/Communism. The destiny of the German people was to withstand these last vestiges of modernity (as Heidegger sees them), what he calls the 'inner greatness' of National Socialism was a response to this mission. Heidegger's mistake was not to subordinate philosophy to a *völkisch* ideology, it was rather to see National Socialists as a response to the crisis which from beginning to end Heidegger understood philosophically.

Philosophically understood, the crisis in which the West found itself was one which was peculiarly unsuited to political action, precisely due to its kairological nature. As he stated in the first Hölderlin course in the winter semester of 1934–5, '[T]he decline [of a people] is … a historically exceptional moment, which can stretch for a century.' (Heidegger, 1989, p. 122) The decline of a people is understood here in terms of its capacity to face its own destiny in relation to the history of being. Such a decline cannot be measured chronologically – as decline it is an exceptional moment, chronologically that moment can last a hundred years. The experience of the decline is an experience of the moment not of the chronologically measurable duration. The concrete experience is one of having been convinced of the self-evident truth of the past order. This past conviction has to be disclosed as illusory. The occasion for such disclosure is generally that of an event which in its horror reveals the illusions of the past order. The response to this horror is often a despairing attempt to find the causes of that event, one which can allow it to be overcome. The seeking after such causes is a seeking for the start of the crisis, an attempt to create a new chronology. But any such chronology covers over the decline because it does not reach or even glimpse the question of the beginning and origins of the decline; it does not question its possibility.

This decline can be understood in terms of its greatness and in terms of the everyday experience of its traces. This indeed reflects a difference which philosophical questioning always encounters:

> When we attempt to ask what philosophy may be, what language may be, what art may be, what the people may be, we always touch on something great within the Dasein of human beings, on something which exceeds and at the same time confuses the individual human being. …Everything great in the Dasein of human beings is at the same time small, both diminished and at the same time ambiguous. In his average everydayness the human being needs this diminishing. (Heidegger, 2009, p. 19; GA 38, p. 22)

Heidegger goes on to say that the mediocrity which goes along with such everydayness is necessary for existence. It only becomes dangerous when it leads to a forgetting of the great. This is in other terms a restating of the relation of authenticity and inauthenticity in *Being and Time*, where the authentic is a modification of the inauthentic. I stress that here because a influential interpretation of Heidegger's politics, Richard Wolin's *The Politics of Being*, argues that the relation of authenticity and inauthenticity in *Being and Time* is one which implies a social hierarchy of the authentic few and the inauthentic many, and that this then allows Heidegger to follow the *Füherprinzip* in 1933 and to understand the German people in terms of a leading elite and the following masses. As he states, 'the de facto separation of human natures into authentic and inauthentic is radically undemocratic. The political philosophy that corresponds to the ontological dualism suggests that human beings are divided by nature into leaders and followers.' (Wolin, 1990, p. 56) In the first place, undemocratic implications of a philosophy can only function as a critical point against that philosophy on the basis of an unphilosophical, that is, unquestioning, commitment to democracy. But, even leaving that point aside, this statement is based on fundamental misunderstandings. As the above quotation from Heidegger should make clear (and at any rate a careful reading of *Being and Time* which is not distracted by Heidegger's occasion slip into culture-critical rhetoric would already show) the 'They' is not a 'term of derision from which no conceivable good can emerge'. (Wolin, 1990, p. 44) As noted above, the term 'they' is misleading, *das Man* is not out there, but rather is constitutive of every self as the everyday interrelation of selves only happens on the level of the 'one': One stands back from the subway doors as they are about to close, one drives on the right-hand side of the road etc. The problem for Heidegger is when authenticity becomes totally immersed in the 'one'. This is a problem fundamentally because the 'one' does not ask the question concerning being. To ask that question is to be called back from the everyday to the place in which such a question can first emerge. The silence of the call of conscience does not at all suggest 'a deliberate infatuation with the forces of unreason' (Wolin, 1990, p. 43), but on the contrary places Dasein for the first time in the position of self-responsibility in which reason becomes truly possible. (cf. Crowell, 2007, pp. 59–60) In fact, what Wolin does not point out is that the call of conscience in common with the occurrence of Angst is that which happens when we least expect it. Far from authenticity and inauthenticity being social categories, they are unpredictable and uncontrollable states of being. The point is to be open to the authentic and not to flee it. This is a decision which every Dasein faces: Against an elitist tradition in philosophy – which one might call 'Platonic' – Heidegger's analysis of Dasein in fact understands the philosophical question as essential for every Dasein. Julian Young is correct to point out that authenticity, far from leading to totalitarianism, is fundamentally opposed to it, actually forbidding commitment to fascism. (cf. Young, 1997, pp. 77–8)

It would not be difficult to read the totalitarian regime as an absolutization of the everyday in the sense that it allows for no call of conscience, would recognize no moment of Angst and is a denial of questioning and philosophy as such. Its displays of 'greatness' function only to tranquilize its populace in their everydayness. Philosophy in general – and Heidegger in particular – can only be implicated in totalitarianism if it abandons or loses sight of its questioning vocation. It does this precisely in its relation to grounds, its impulse to found and establish. It is then vital to return to Heidegger's understanding both of the crisis of 1933 and the appropriate philosophical response to it.

The decline contains within itself the possibility of revolution. In the face of this situation both pessimism and optimism are misplaced: 'the childlike categories of pessimism and optimism have long become absurd.' (Heidegger, 1987, p. 38; EiM, p. 29; cf. Heidegger, 1994, p. 50; GA 45, p. 55) What is meant is that the decline of the previously existing order is such that every expectation – whether hopeful or hopeless – that what exists could still (or also not) be saved, misunderstands the situation. The situation is one in which a revolution is called for in order to overcome decline and dissolution. Resoluteness is empty because it is a preparedness to respond to the unconcealment of being, a readiness which is not to entities as they were, but to the possibility of entities to come. The moment of this readiness is not chronologically determinable and what follows from this is that the time of revolution is *radically indeterminate*. Even when everything may appear to indicate its coming, there is no form of knowledge – understood as technē – of a revolution. Knowledge which orientates itself on the work, which orientates itself on founding, which aims to establish on the basis of a certain openness a new order of entities, fails in the moment of revolution where the old order no longer offers any guidance and the new order is still to come. This is a moment where the decision to repeat the having-been opens up its possibilities, and these possibilities give space for action. But philosophy, in asking the question of grounds, is a questioning not aimed towards those possibilities, but rather is directed at the unconcealment of being itself which allows those possibilities to be.

In 1933, Heidegger made a decision for a certain ontic possibility amongst others; namely, for Hitler and the National Socialists. It is hard to miss the difference in tone in Heidegger's writing in this period and not only in his 'political' speeches. The Rectoral Address is replete with voluntaristic overtones. This is true neither of his writings and lectures prior to 1933, nor after 1934. This voluntaristic stridency is combined with certain Platonic motifs in particular with respect to the threefold division of labour, armed and knowledge service. What unites these two elements is a forceful decision to think philosophy as founding. Despite the fact that Heidegger stresses the 'question-worthiness of being' (Heidegger, 1985a, p. 477; R, p. 16), the address clearly places philosophy in a founding role, which is confirmed by the manner in which he sets in other text the thinker alongside the poet and statesman as the 'founding fathers' of the new state. As Pöggeler very well shows, Heidegger was motivated by the ambition, if not the hubris, of leading the *Führer – den Führer führen*. (Pöggeler, 1985) But this ambition betrays a basic insight of *Being and Time* into the nature of possibility.

Hans Sluga has seen the issue here when he states that Heidegger was guided by a fundamental misunderstanding of the relation of philosophy and politics, namely that politics could be given a philosophical basis or foundation. (cf. Sluga, 1993, pp. 230–3, 245–53) In this, Sluga maintains that Heidegger went against the radicality of his own understanding of philosophy. If philosophy is for Heidegger a 'pushing across all boundaries into what seems at times a black hole of nothingness', then he should have seen that 'he would have to abandon the claims of thinking to be a founding authority and instead accept its giddy freedom.' (Sluga, 1993, pp. 231–2) This critique is in essence correct, but what Sluga does not sufficiently recognize is the conflict in Heidegger's own thought between an understanding of philosophy as questioning and philosophy as a thought work (*Denkwerk*). The latter understanding of philosophy gives to it a leading political role and as such it is the case, as Wolin argues, that in

Heidegger's work there is a 'politics of being'. But then, this is true of a certain way of understanding philosophy which goes back to Plato.

The city state which Socrates describes in the *Republic* is not a real state, but a poetic invention. It is as such the product of a technē. The polis is thus a work, and the role of the philosopher is to care for this work as a whole. On closer inspection, however, the philosopher has in fact a double role: on the one hand, he creates the work of the polis, insofar as in a utopian poiesis he reconstructs the polis, but on the other hand, he plays a specific role within this work. This latter role is that of ruler. He must rule the state for the same reason as he has the task of creating it, because he has sight for the ideas and as such knows the paradigm of the state (Plato, *Republic*, 592a8–b5) The city as a work makes possible the praxis of its citizens, but this praxis if uncontrolled can lead to its destruction. Thus the philosopher needs to ensure that the praxis of its citizens is in line with the order of the city. It is this which leads the philosopher into an ambiguous position, which Plato expresses by saying that the philosopher in essence does not wish to be a ruler, but is forced to be one. (Plato, *Republic*, 500d4) The philosopher has the desire to pose questions, but in order to satisfy this desire he must construct and rule a city which is set up according to its ideal. In a city which the philosopher does not rule even praiseworthy qualities are causes of corruption. (Plato, *Republic*, 491b4) Hence, the ambiguous relation of philosophy to politics.

This ambiguity is to be found in Heidegger during his time as Rector. In his Rectoral Address, he asks whether there should be science at all. (Heidegger, 1985a, p. 471; R, p. 10) He does not give an immediate or direct answer to this question, but it is clear that this is a question which concerns the fate of his listeners: this is so because knowledge – in the sense of technē – 'remains as before delivered over to overpowering fate and fails before it.' (Heidegger, 1985a, p. 472; R, p. 11) Technē is not to be understood here as the application of a model on receptive matter, but rather remains delivered over to the power of the unconcealment of being. This understanding of knowledge stands at the beginning of philosophy, according to Heidegger. This does not amount to a definition of knowledge, but rather brings to expression a certain *experience* of knowledge. This experience conditions the destiny of the west. The beginning, however, remains before us. (Heidegger, 1985a, p. 473; R, p. 13) That means: the Greek experience of being remains hidden. Human beings presently live in the decline of that event of the originary Greek experience. The question then for the philosopher is how it is possible to act in the moment of that decline in such a way that would be fitting for an other, a new, beginning. This is to say, the philosopher needs to find a way to live in a revolution which must erect its own measure of a beginning. This beginning

> has invaded our future. There it awaits us a distant command bidding us to catch up with its greatness. Only if we resolutely submit to this distant command to recapture the greatness of the beginning, will science become the innermost necessity of our Dasein. (Heidegger, 1985a, p. 473; R, p. 13)

To think towards the new beginning requires a repeating of the first beginning, but this in no ways means a subordination to it. A common but fundamental misunderstanding of Heidegger's account of historicity is one which understands the decision of submission as one of subordination. (cf. for example Harries, 1978, pp. 320–1 and Wolin, 1990, pp. 64–5; cf. for a critique of this interpretation, Young, 1997, pp. 84–6) The choice of heritage in *Being and Time* – which is alluded to here – is not the choice of a actuality but of a possibility. It is to repeat that heritage and to transform it in the repetition. So too here: the Greek world cannot be reinstated, it is precisely due to the exhaustion of that world that a new beginning is made necessary. Heidegger claims this new beginning to be a German one, but the Germans are in essential opposition to the Greeks, the German repetition of the Greeks is a transformative one. (cf. Heidegger, 1989, pp. 290–4) Indeed, as Heidegger will later make clear in the *Contributions to Philosophy*, the mood in which the transformation which comes with this repetition is disclosed is one of terror (*Schrecken*) rather than wonder. It is a terror in which the fundamental unknowing and uncertainty of the situation is disclosed. (Heidegger, 1999, p. 11; GA 65, p. 15; cf. also Held, 1991, pp. 40–1) Set as we are in this mood before the question-worthiness of being, there can be no setting-in-work. The situation is precisely one for the thinker in which nothing has been brought to a stand. In such a situation, the temporality of the revolution itself remains in question. This Heidegger himself admits, 'we have no right to suppose that the elucidation and unfolding of the essence of the German university could take place in the current or coming semester.' (Heidegger, 1985a, p. 478; R, p. 18) In such a situation the philosopher as questioner strives to keep the question-worthiness of being open. But where is there a trace of such questioning or openness in the apodictically stated claim: 'Precepts and ideas are not the rules of your being. The Führer is the one and only present and future German reality and its law.' (quoted in Pöggeler, 1993, p. 214) As Pöggeler points out, Heidegger is in effect playing Hitler against his party here: it is not the party programme of the Nazis, but rather the creative will of the Führer which will lead Germany (and as Germany is to lead the West,

Europe as a whole) through the coming revolution. (Pöggeler, 1993, pp. 214–15, 240n. 7) But how can Heidegger know that the 'Führer' is really the leader, and indeed, why should we think that a leader is needed at all (at least in the sense of leadership claimed for and by Hitler)? The movement of the kairological situation reveals the contingency and freedom of being; Heidegger knew this as the questioning philosopher, but as ruler (of the university) or as educator of the ruler, he did not wish to see it.

But precisely because Heidegger has here fallen victim to the ambiguity of philosophy, it is not as simple a matter as merely accusing him of blindness. On the basis of the experience of grounding, philosophy sets the most radical of questions, the why question. The philosophical question and the problem or task of grounding are closely connected. The relation of philosophical questioning and grounding can be understood in temporal terms: only on the assumption of a chronological, i.e. causal, continuity can something be grounded. At the same time, the experience of grounding is one of the limits of this continuity and the opening up of the abyss, the absence of grounds. Questioning finds itself before this abyss constantly, as it sets in question the order of entities and as such seeks the source of all order: being. Being can, however, only be 'experienced' in the midst of the moment of vision – the kairos – and then as transformation, as revolution. This changeover from chronos to kairos makes philosophy ambiguous in the sense that in the experience of grounding it turns to questioning, but in the fear of the new, and thereby its apparent political powerlessness, philosophy gives up its questioning and places itself at the service of the exercise of grounding. When Heidegger claimed for the philosopher a ruling role (as educator of the Führer), he gave up on the philosophical calling to question.

For Heidegger, the thinker plays the leading political role because he mediates between the poet and the statesman and in that sense engages in politics in the 'highest and most authentic sense'. (cf. Heidegger, 1989, p. 214) The philosopher has no eternal model to guide him so must direct his attention to the situation, in order to actualize the new beginning. That which is actualized is the thought-work which arises in relation to the artwork of the poet and the state-work of the politician, all of which found the polis-work. Only when these three works stand in relation can there be a true, that is, spiritual – *geistige* – revolution. (Heidegger, 1985a, pp. 474–5; R, p. 14; cf. Derrida, 1989, pp. 31–46)

The question, however, is whether a revolution can be 'actualized' in this sense. Revolution is glimpsed as possibility in the decline. I have already quoted Heidegger on the nature of decline, but it is worth quoting again, this time in full:

> The decline [of a people] is for that reason a historically exceptional moment, which can stretch for a century, because the inexhausted, the inexhaustible of the new beginning, the *possible,* can bring itself there to power, but only if those are there who are capable in advance of experiencing, of founding, and of knowing and of effecting this inexhaustible possible as such.' (Heidegger, 1989, p. 122)

The new beginning is the inexhaustible possible which can bring itself to power. There is no reference here yet of any creator, certainly not of a philosopher. The new beginning is a power which is capable from itself of bringing a new order on entities. Such a power it is not actual, but is rather the possible. There is here no talk of actualization; the possible does not stand in any chronological time. The indeterminacy of this possible is that of the decline, as the possible of the new beginning is nothing other that the possibility lying in the decline. Because of its chronological indeterminacy the decline is a 'historically exceptional moment'. The new beginning as possibility is transformation in which all entities find themselves ordered under a new power. Being, however, needs human beings. It needs those 'who are capable in advance of experiencing, of founding, and of knowing and of effecting this inexhaustible possible as such.' In this list we can, without difficulty, recognize the thinker (who knows) the poet (who founds) and presumably the politician (who effects). But the new beginning is first experienced, and is experienced as the possible. This experience is that of questioning. Only then does experience becomes 'knowing, founding and effecting'.

All of this suggests that the human being, and above all the philosopher, possesses a peculiar *unknowing* in the face of the revolution. This unknowing allows her to be prepared for the revolution, but not to bring it to actuality. The kairos irrupts as an alien power. Suddenly, abruptly, the world is otherwise than it was; the order of the polis is transformed. The setting-in-work does not constitute the revolution. To the extent to which polis was an order of entities it was more than a work. It is this 'more' which indicates the place of praxis and this is disclosed above all in friendship. Revolution shares with friendship a manner of being as a between-event. In both, the mediation of work falls away and human beings are thrown back into a fragile intimacy. Aristotle caught a glimpse of this when he said that friendship both held the polis together and did not require justice. (Aristotle, *Nicomachean Ethics*, 1155a24 and a27) Friendly action in the between-space of possibility transcends the work in the direction of praxis, which is without knowledge (technē), but has a sense for the new.

V The chronology of revolution

The word 'revolution' is one which we associate above all with political events and not without reason: a revolution as a transformation of order is a change in the relations of human beings with one another. Heidegger's ontological understanding of history is not a denial of the political, but rather an attempt to understand the political in the context of a fundamental revolution in the way entities appear to us, a transformation with direct political aspects and consequences. Heidegger did not look to the revolutions of modernity to exemplify his thinking, because presumably these did not involve a radical reconstitution of the order of entities, but rather 'remain within the overarching sway of [the Greek beginning's] claim that posits the truth of beings in terms of the correctness of human thought and representation.' (McNeill, 206, p. 126) Nevertheless, as already pointed out, the very idea of revolution in this transformative sense is a modern one. It is revealing the extent to which revolutions are associated with particular days, such as the 14 July 1789 or October 1917. These dates are remembered because a world irrupted in them which did not so much lie in the aims of those involved in them, but rather was not imaginable before them. All actions embody certain intentions which assume chronological expectations and the corresponding responsibilities. And when a new order emerges those involved may not know how to deal with it, but they do know that it did not fall from the heavens, and came to be in their actions, actions for which they have a shared responsibility. The new which arises in the moment is strange, but is not without its history. It comes to be in action, but was not present before. This absence forms a veil between present and past. Only with difficulty, but then never totally, can this veil be lifted. The effort to do this, however, cannot be avoided, above all in political action which requires reflection. (cf. Held, 1993, pp. 407–8) Action in an epochal moment while not reducible to chronology, to chronological reflection and responsibility, cannot at the same time transcend these. If it seeks to do so it lays itself open to fundamental error. In the case of Heidegger, it is impossible to disagree with Pöggeler when he states that 'clearly [Heidegger] lacked the minimum of necessary political reflection'. (Pöggeler, 1990, p. 34)

Conclusion

This book set itself the task of investigating the split in time between kairos and chronos with and also against Heidegger. By way of conclusion, I would like here to 'repeat' the ways taken with a view to bringing out above all else the manner in which kairos and chronos cross over each other and in doing so to try to bring to light the temporal structure of revolutionary time as a the time of a historical event. The way this is understood has consequences in the manner in which we read the later Heidegger beginning with the *Contributions to Philosophy*. These considerations regarding kairos and chronos and the time of revolution also has resonances beyond Heidegger in the work of Benjamin and Adorno, the later Merleau-Ponty and Levinas, Derrida and Foucault, Badiou and Agamben. I can at most point towards those echoes here. The task I set myself is a more modest one, namely, to outline the crossing over of chronos and kairos, praxis and poiesis.

As already alluded to in Chapters 4 and 5, Heidegger replaced the concepts of philosophy and philosopher with those of thinking and thinker in the 1930's. Two decades later in a lecture with the revealing title 'The End of Philosophy and the Task of Thinking', he discussed this question in the context of the 'attempt undertaken again and again ever since 1930 to shape the question of *Being and Time* in a more primordial fashion.' (Heidegger, 1977a, p. 373; ZSD, p. 61) We must read this in the light of his statement in 'The Letter on "Humanism"' that the language of *Being and Time* was still 'metaphysical'. (Heidegger, 1998a, p. 250; Wm, p. 325) As concerns the present work the question which needs to detain us is whether this difference of thinking to philosophy has any relevance with respect to the question of the chronos and kairos and the time of revolution.

That which hinders philosophy in understanding the crossing over of chronos and kairos is the ambiguity, discussed in the last two chapters, between the will to ground and dependence on poetry. Since Plato, however, philosophy has never accepted this dependence and understood itself from the experience of technē, hence obscuring the distinction between praxis and poiesis. Furthermore, if time is understood on the basis of poiesis, then the kairos is hidden and the relation of time and eternity is conceived as at once a mimetic relation and a product of decline. In contrast, Heidegger comes to see thinking as that which attempts to secure a liberation from technē. The task of thinking has, he later says, 'only … a preparatory, not a founding character.' (Heidegger, 1977a, p. 378; ZSD, p, 66) As such thinking does not set itself in opposition to poetry, but recognizes that thinking and poetizing are two distinct ways to approach the moment of vision, the kairos. If this is so, then the distinction of chronos and kairos can no longer (as it does in *Being and Time*) complement that of poiesis and praxis, but rather intersect at a cross angle with it.

Thinking, which has no intention to found, is not that of a creator; it is a thinking which holds itself ready and which tarries questioningly. This place of questioning is the most difficult to sustain because it is a thinking neither of chronos nor or kairos, but of the difference, the crossing over, of both. There is a chronological conditioning of thinking, namely its belonging to a continuity. Thinking, as Heidegger tells us, is rooted in the capacity to listen. Only that capacity to listen allows us to be in conversation. Our thinking and our speech does not happen in isolation, but rather in conversation with others – be those others texts or works of art or jugs or trees or waves or people. In each case, in listening to an other, I am listening to an other with its, his or her own chronology or rather chronologies, which both bind and separate us. Thinking responds to the claims of another, to that claim which is the actuality of the other, its stretching out over a certain, distinct chronology and as such responds to the other as it appears in the everyday. The other comes into play for thinking when it asks about being and as such about the making possible of the order of entities. However, that order is as much an order which is lived and which is manifest in continuity, in a continuation of that order, in which the different continuities of all those others I encounter there have their own sense and their own identities. Nor is this identity a fixed one: as Ricoeur shows such chronological continuity depends on its constant reproduction through narration. (Ricoeur, 1984, pp. 70–7)

The time of narration is not simply the time of a conversation with myself, I always narrate something to somebody. Narration is dependent on a common horizon of experience with the other. The narrative time of my narrating has within it the tendency to order itself within the wider story of a community or ultimately of humanity or indeed of all worldly entities. In Chapter 2, we spoke of the erotic kairology of this between-time in friendship, but normally the between-time has a chronological character and through this chronology makes narrative possible. Narration is constituted in and through poietic time; narrative identity is poietic. It always relates itself back to the past: the narrator allows a past to arise which lies 'behind' her. Narration opens up a future also;

we narrate in the expectation of a still open future. But we cannot narrate the future, at most we can narrate a vision of a projected future. As a certain point we must forgo narration in order to face the risk of the future. The uncertainty of the future threatens everyday chronology as well: we are always at risk that the continuity of our lives will be torn asunder by some future event. All those things that I expect in the future will continue to be my parts of my narrative – my career, my home, my family – can be lost all at once.

In such an instance, when the everyday certainty of continuity breaks down, the human being is placed in a totally new situation. In one moment, the whole of a life can appear otherwise; the expected chronological continuation of past relations is closed off. The past 'essences' still in the present, but it has become alien and the continuity of that which is one's own is placed in question. Decisive, however, is that the continuity of the chronological rhythm penetrates the kairos itself. Chronological rhythm is the rhythm of habit, habituation and custom which ties us to a past which 'essences' in the present as a past carved into our very flesh. Such habituation binds not only past and present, but also my chronology with those of others. It is the formation of ways of acting which form the character not only of individuals, but also of communities and peoples. Moral norms themselves are based on such habits, or, as Bergson puts it, on the habit of forming habits. (Bergson, 1963, p. 26). Such habits, and such a habit to form habits, are constitutive of chronological time and reach right down to bodily movement in its gesturality.

Habits mould the singularity of the person. They are masks (*persona*), which, however, are not distinct from the 'true' face. These are moving masks, they mould us in our flesh and blood. They are so much part of us that the new horrifies us if it threatens to dissolve these masks and destroy past habits and past customs, even when the new opens up liberating possibilities. The time of revolution is the time in which the rhythm of such habits is interrupted. The habits and customs of the past which have been so familiar and so trusted lose their certainty, and in the loss of their rhythm they are opened up to analysis as if for the first time. The habits remain but the loss of rhythm is a loss of certainty in their guidance and correlatively the suspicion which they evoke in others. Those customs which used to bind a community now become signs of an unhealthy commitment to a lost past and as such objects of suspicion, condemnation and ultimately denunciation. They speak of an implicit justification in terms of moral norms, which in the time of revolution are unjustifiable, reactionary. On the other hand, habits and customs are not simply mechanical processes – although they can become such – but rather good habits are continually being affirmed and renewed on the basis of their presumed goodness. That presumption has many layers from the conceptual to the bodily and can appear in many spaces from the public to the private. The interruption of the rhythm of habit while making problematic the self-evidence of habit can no more destroy it completely than anyone in the time of revolution can leave behind their own bodies.

Already in Chapter 1 the question was posed as to whether action in the kairos does an injustice against the past to the extent to which it arises from a forgetting of the past. We can now make that question more precise: is kairological action against the other to the extent to which it departs from all past commitments and obligations? If kairological action is exclusively kairological, then the answer would have to be yes. But that is to assume that the kairos can depart entirely from the chronos, that the time of revolution is a total break with the past.

Thinking prepares for the kairos or better it holds itself in readiness for it. Thinking repeats the having-been, in order to hear the 'unsaid' in it. (Heidegger, 1998a, p. 155; Wm, p. 201) It thinks the said but attempts in a violent manner – Heidegger speaks of the 'justified violence' of 'philosophical dialogue' (c.f. Heidegger, 1990, p. xviii; KPM, p. xvii) – to get behind the said in order to come to that which lies unthought within it. The thought of past philosophers has its rights, but only through that which its holds ready as possibilities of thinking, not what is fixed doctrinally. Again we see how Heidegger's account of past and having-been, actuality and possibility, continues to shape his thinking: thinking listens to the having-been possibility and does so in a violent surpassing of the past actuality of a philosophical doctrine. But something in the actual resists this 'repetition'. This resistance marks an alterity which lies outside all repetition, it remains a voice against such repetition and ultimately against kairological temporality itself. It is a voice which calls for the promises and the habits of the past to be respected in their own actual validity. In the kairos, thinking does listen to the other; but in this hearing the other is encountered as a stranger. The work, which binds, loses its power; the masks, which were familiar, have fallen. Levinas speaks here of the 'naked face' of the other; we might remember here the account of equality in the moment of revolution which was discussed earlier. That naked face is the face which offers no guide as to what is expected of me, which corresponds to no custom, habit or moral norm. This is a face which calls on me to respond, while leaving me totally responsible for how I can respond. This is a world which has become alien.

Nevertheless, a world which becomes alien in the kairos has *become* alien. A thinking which wishes to think the kairos ought not avoid the question *how* the event of this alienation happened. The latter question expresses responsibility directed at the past, seeking a chronology and a reconstruction of ties through a new appropriation of the past. This does not mean that it seeks a causal explanation; rather, it inquires concerning the chronological context in which the kairos irrupted.

Heidegger recognizes one form of obligation, namely that to a people. This obligation, expressed in the pronoun 'we' is an obligation to a place, a home, in which a thinker finds himself. That belonging marks a relation of obligation and commitment. Although in Heidegger's engagement with the Nazis this obligation took on a dangerously exclusive form, his more reflective discussions make clear that the obligation to this past is an obligation to home which precisely needs to be decided upon. Such decision is one which renews a commitment to others, a commitment to a shared chronology, but one which is not unquestioned, but precisely is up for question in the kairos.

The voice of the other is perceivable both in chronology and in kairology. The thinker bears a responsibility towards it. This responsibility towards the other entwines self and other, thinker and that of which she thinks, in relation to promising: in promising to another the self is bound to that other. Her place in the history of the other is that of the second person: 'at that time *you* were …', '*you* promised me then …'. In the story of another's life, the self's own Dasein remains always an alien element, precisely to the extent to which that self co-determines this story and in this way bears responsibility for it. It is only to the extent that the self can ignore or forget this participation in the story of another that she is in the position to leave chronology behind. This account of the time of the self in the chronology of another should not be confused with 'between-time'. 'Between-time' is neither mine nor yours but a power between us. Such time plays an essential role in promising. In promising to an other, I and she are subject to the power of the time between us, the power which binds us together. But the promise while opening up possibility is itself an actuality which binds me to her chronology (and she to mine) as a stranger. The strangeness, the alienness, consists in the fact that her chronology is conditioned by another chronology which is not commensurate with it. This alienness resists all repetition because I do not have any direct access to it. In the between-time I encounter the chronology of the other as an alien moment in my own, the moment of her promising which binds her to me – an alienness which displays the fragility of the between time.

The self bears responsibility for her role as stranger in the time of another. In this there is an obligation regarding the reality of the past, i.e. of that time which is not repeated in the kairos, but which ought to be respected. The kairological situation cannot, however, be understood without reference to chronology. This is so because such a situation arises out of action – action with and amongst human beings. The situation is not created by anyone or a collective of people in the sense that it results from kairological intentions: the latter is a contradiction in terms, not because nobody ever aimed to bring about a revolution, but because such aims are always limited to specific goals, which always assume a given horizon of action. Nonetheless, those in the kairological situation are responsible for it, responsible not as creators, but as those for whom and in whose actions time becomes other: they are responsible as being already guilty. But what is the nature of this responsibility and how can obligation be understood in that context?

The key terms here is the 'becoming-other' of time. The way in which the power of time is experienced, namely in rhythm, gives us an indication of what is involved here. In chronological time we have the capacity to recognize the right time. We can see this in such diverse fields as politics, economics, rhetoric, theatre and sport. In each case, although an element of calculation is necessary, it is not sufficient. In order to recognize the right time, a sense for the rhythm of the respective times is needed. This sense assumes a between-time: only in a situation in common with others, only when time is not simply mine or the others does the self need such a sense. This sense can at once fall into a rhythm of mechanical monotony. But it can also be a sense for the becoming other of time, for the failure of routine and the change of rhythm.

In such a situation, a well-honed sense tells only that the situation is otherwise, that routine measures are inappropriate. But those in that situation are at first clueless as to what can or ought to be done. The situation has changed, but what is to be done? As Heidegger makes clear in his interpretation of Aristotle, movement means becoming-other. Mere routine is characterized by a levelled off form of such becoming-other: for example, the transition from the now of meal time to the now of work. There is no break in such becoming-other, on the contrary it grounds – as Aristotle also recognized – the constancy of time. However, without such becoming-other of chronological time there would be no possibility of the kairos. The rupture of the kairos is possible only due to the fragility of chronos, which is manifest in the movement of becoming-other. This does not mean that kairos is simply an extreme case of such chronological becoming-other. There is in becoming-other a tendency to dispersal, such that it is to a large degree the work of narration to counteract this tendency. The kairological situation on the other hand is characterized by the unity of temporal modi. As already discussed, such dispersal assumes a temporal order. But in the becoming-other, a transitoriness can be experienced and precisely this is the seed of the collapse of order itself. The kairos is not so much the most extreme form of becoming-other, but rather is to found at the core of the becoming other itself, namely in the chaotic juncture of transition. In the movement of transition, there is the possibility of transcending chronology. This chaotic transitoriness cannot be immediately experienced, but is rather indirectly sensed in the fragility which is in every rhythm.

The kairological situation is a qualitative difference in time in which the continuity of transition is overturned into the chaos of transitoriness itself, whereby the becoming-other does not confirm constancy but undermines it. The harmony of time's rhythm is overturned into moment of pure transitoriness, in which time as the power of arising becomes manifest. In such a moment the human being is without knowledge (technē), but he retains a cognition of rhythm. Although he does not know how to act, he may still recognize the rhythm of this time, even though it is not the rhythm of chronological time. The sense for this rhythm allows the human being to act in the kairos, to act in and with a revolutionary situation, in which as we have already noted the rhythm of time may so quicken up as to allow for no reflection. But the question arises whether acting with the rhythm of time is a matter of *cleverness*, rather than goodness. It may be, for example, that Heidegger acted with the rhythm of the opportune moment in 1933, but if this acting were mere opportuninism, merely acting to succeed in that moment, then the kairos loses all ethical sense and the chaos of the moment allows for terror and violence.

To make a distinction between acting in the opportune moment and acting opportunistically, Heidegger's account of preparedness is of crucial importance. As is clear from the place preparedness plays in Heidegger's account of conscience, thinking as preparedness for the kairos is a preparedness to act. Such preparedness is not the preparing of certain principles, to which Dasein must hold fast in the kairos, because such principles are promises through which the self remains tied to its alien place in the chronology of another. Rather, preparedness for the kairos is a readiness to act in terms of a sense for the trace of the other in the between-time. This sense is, above all, a feeling for rhythm. A responsiveness for the other arises out of this feeling. The erotic relation to the other shows in an exemplary manner how we can learn to move in the between-time and to become receptive to the motion of an other. Responsibility announces itself in this responsiveness. Only out of such responsiveness for an other, can there still be responsibility when those mediations of obligation break down. In such a situation there occurs a peculiar alienation in which the self recognizes itself as a stranger to the other, recognizes in other words, that its own being is radically severed from the 'one' [das Man] and that its being with an other is possible when social masks fall away. The self sees itself in this between-time as having always been the alien second person in the chronology of each other.

Responsiveness so understood is a feeling for the rhythm of the chronology of an other. But this responsiveness, although one which sees itself as alien to the other and hence which can feel its own second person status for the other, is that which guides the self in the between-time in which it encounters the other without mediation. The kairos shows the other as other and opens up an ethical space in which the other is encountered in its possibility, i.e. as escaping all orders of past and future, and as a chaotic moment of having-been and to come. In this occurrence the happening of being and the epiphany of the other – epiphany because appearing as beyond chronology – are one and the same. The destiny of being is the opening up in the moment of a new possibility for entities, and this arising of the new happens in the between-time, it is experienced as the power of time in the rhythm of unmediated relations.

But the chaos of this moment is a great risk and the experience of the other in this moment must remain tied to a certain chronology. The arising of time and the appearance of the other as possibility cannot obscure the passing of time and the actual presence of the other. In rhythm time arises, but it passes also; the other appears as possibility, but has also actuality. In acting in the kairos the self is indebted to the actual as well as the possible, and is subject to the passing of time as well its emerging.

Time is historical to the extent to which history is not made but happens, and happens in the moment of vision. The philosophical problem of history for Heidegger consists not so much in the making of continuity through narration as in thinking discontinuity as discontinuity. 'History' not only as historiography, but also as the stories we tell of the past, hides these discontinuities in composing them, yet discontinuity is constitutive of historicity. In the lecture 'The End of Philosophy and the Task of Thinking', we read that thinking is 'forced to think the historicity of that which grants a possible history to philosophy'. (Heidegger, 1993, p. 378; ZSD, p. 66) Thinking does not act to found a history, but to think the *possibility* of history. Thinking thinks the moment, in which history arises. Such history is not the necessary consequence of the moment, but rather the possible history which has its origin in the historicity of the moment. Heidegger speaks throughout that lecture of the necessity of waiting and the preparedness of thinking for that which, although hidden in the beginning of philosophy, may come to appearance in an unreckonable time. This understanding of historicity is a kairological one, in which the possibility of history lies in discontinuity.

'History', according to Augustine (and in this he gives the basis for the Western understanding of history at least until Hegel), is a concept which makes the uniqueness of the birth of Christ understandable, a uniqueness which cannot be thought in the Ancient cyclical account of human time. (cf. Augustine, *The City of God*, book 12, 14) This history is then generally understood as a teleological chronology, which receives its significance and meaning from the birth and death of Christ – a history of redemption. The kairos is claimed as a ground for a particular history. The being of this ground is understood in terms of justification. The eschatological hope has its ground – it justifies itself – in the appearance of the eternal in time. But the question arises as to how history can receive its meaning and significance from a kairological event which is precisely unique. The question here is how the uniqueness of the kairos can relate to particular chronologies which look to it for justification and those chronologies (in this case the sometimes violent interpretation of the Hebrew scriptures) which seek to justify retrospectively the kairos itself.

To approach this question it is important to return to the principle of order in *Being and Time* discussed in Chapter 1. In terms of that principle there should be a foundational order, that of the Temporality (*Temporalität*) of being. However, Heidegger abandoned this attempt. This can be traced to the failure of the ordering principle itself. Being cannot be the last principle of a foundational order because it does not have the character of grounds. Being, as Heidegger in the wake of *Being and Time* gradually comes to see, happens as the historical in history. The occurrence of this, the event of being (*Ereignis*) plays itself out in the tension of chronos and kairos. The crossing over of chronos and kairos forms the place of the question of being. When the movement of being only comes to appearance in its difference from entities, and when the crossing over of chronos and kairos forms the place of this coming to appearance, then the truth of being happens only in the interrelation of the history of being and of ontical history. The moment of vision, the kairos, can only be adequately thought in this interrelation.

That which Heidegger understands as a foundational relation in *Being and Time* is more genuinely understood as an interrelation. It is historical insofar as in it the conditioning and the conditioned belong together in one and the same occurrence and are reciprocally changed. The history of being is the history of this occurrence as this occurrence takes place in a particular ontical history. Kairos does not occur without chronos, but is irreducible to chronos. The uniqueness of the kairos is only in its difference from chronos, but that difference is only expressible subsequently, and is expressed only in the gaps, the interruption, the unsaid between the words which attempt to account for it. Before this there is only the speechlessness of fundamental mood. 'Angst robs us of speech'. (Heidegger, 1998a, p. 89; Wm, p. 113) When we afterwards in the 'clearness of vision' attempt to account for this moment, then 'we must say that in the face of which and concerning which we had Angst was "properly" – nothing.' (Heidegger, 1998a, p. 89; Wm, p. 113) The speechlessness of fundamental mood discloses the chaotic nature of the kairos. The unique lies outside every context, without cause and without effect. It is this which is the chaotic in it and which gives it an erotic sense. In Plato's *Symposium*, when Alciabiades speaks of his love for Socrates, it is the uniqueness of Socrates above all which evokes his love. (Plato, *Symposium*, 221c3) Love requires no further justification; on the contrary, when it seeks grounds it is already lost. Love can find no basis in the 'whatness' of the beloved, because everything which can be accounted for in those terms are general qualities which the beloved shares with others. In the same way, the uniqueness of the kairos cannot be captured in with the determinedness of a 'what'. As such, the kairos escapes all ordering principles, which order with respect of grounding and general categories, and as such is chaotic.

But the unique occurrence calls to be made perceivable in the actuality of a work. The motion of the kairos, the chaotic within it, is strictly speaking not perceivable or experiencible because experience and perception are already ordering principles. This motion is experiencible only in the traces it leaves at the boundaries of experience, in those chaotic moments in which the ordering movement of experience is interrupted, dissolved, thrown open by that which refuses it.

When someone acts in a kairological moment, she does so unknowingly, but a new chronology begins with that action. In the rupture a new history begins, something new emerges. The subequentiality which characterizes the experience of the kairos, refers to the moment which in the very attempt to grasp it, eludes such grasp. The rupture is experienced as intensity (as we saw in Chapter 3) and this intensity is the trace of the arising of time in the moment. The action responds to the kairos, which is sensed in this intensity; but this response, as the formation of an answer, is directed towards the founding of a new chronology. The response to the unique undermines the striving to return to it; it always amounts to a coming to terms with the kairos. This coming to terms is what occurs in the setting in work of kairological truth, whereby the kairos is taken up poietically. In poiesis, but also in praxis, action in the kairos is concretely an action in the crossing over of kairos and chronos, in which kairos is experienced as intensity and rhythm in the chaotic moment of unknowing response.

The human capacity to start is always a response to the beginning that addresses her. The possibility of such a start lies in birth. Between Chapters 1 and 4 the notions of possibility and of birth were modified with the shift from a question concerning the historicity of temporal Dasein to the question of historicity as such. As we saw, historicity cannot be founded on the temporality of Dasein because the possibility of Dasein is itself a historical possibility occurring in the event of being. Within this lies the openness of Dasein for transformation. Such transformation can be understood not as renewal (birth), but as metamorphosis. Metamorphosis remains alien to Dasein in its natality and mortality; it withdraws from all experience and all conceptuality. Nonetheless, within natality itself there is an abyss in the chaotic moment of birth between the womb and the world, in which a child without a name exists as a stranger in the world. This moment of alien strangeness is sensed in the human striving for home, whereby the world remains always unhomely, as we say uncanny, literally a place in which we do not know how to find our way. This strangeness is disclosed in kairos. The world is historical for the human being, not because he has the capacity to appropriate the past and to form identity in that way, but because in every such attempt the strangeness of his birth and his place in the world is disclosed in that which withdraws from such attempts at appropriation and identification. With respect to birth, this means that it is not the capacity to begin, but rather the possibility to respond to a beginning, which remains.

The rupture with chronology remains thus outside every chronology. The occurrence of the new in chronology is only experiencible in its strangeness, its resistance to appropriation. There is here a double movement: towards the kairos, which always fails, and from the kairos, which happens only with profound loss.

The crossing over of kairos and chronos is chronologically determinable to a certain point. Kairos occurs in a chronology. While the kairos is not caused by prior events or circumstances, we can still say that it is occasioned by them. The destiny of being occurs in entities. It is an occurrence, which cannot be divorced from ontic history. The kairos is characteristic of a time and of a certain history; it is occasioned by that history and refigures that history. The time of revolution is one in which the past and future is transformed and a new chronology becomes possible, such that the past can no longer be thought except in relation to that revolution, to that kairos.

The kairos is without ground and does not function as a ground for a chronology. History has no ground either in the temporality of Dasein or in the destiny of being. In the kairos, the origins of history are sensed, i.e. an initiating movement, which is brought to a stand in it. Historical time is not a united order, but rather is a tension of chronos and kairos. In questioning, thinking seeks to return to the origin, strives to think the discontinuity of kairos. In this sense, the task of thinking does not lie in conceiving a philosophy of history, but rather in understanding historicity, hence the chaotic uniqueness, which cannot be sublated into history, but nonetheless allows itself to be sensed.

The task of thinking then is to think the possibility of history. As poetry attempts to bring the unique to language, we can say that poetry is closer to thought than historiography. In this way we come in the vicinity of Aristotle's thesis that 'poetry is something more philosophical … than history' (Aristotle, *Poetics*, 1451b5), admittedly for almost the opposite reason. Both poetizing and thinking relate to the kairos as a unique event, both attempt to bring the singular and unique to language.

As I stressed from the beginning, the kairos is ambiguous, it is a time of transformation for good and for evil, for victory and defeat, for the rise of the new and the demise of the old. It is a time without guarantees. For St. Paul the kairos is the time of redemption, it is a promise, a covenant. At the same time the kairos, as we have seen, endangers the very continuity which promising appears to demand. As such, the kairos cannot be understood without qualification on the basis of promising. In its Homeric sense, the kairos is at once the bringer of good fortune to the one who strikes it and misfortune for the one whose vulnerability is thereby uncovered. Understood in temporal terms, this points not just to an ambivalence, but also to an unknowing; we do not know whether it will bring good fortune or misfortune, redemption or catastrophe.

In the light of this, we can hear the lines of Hölderlin's poem Patmos, which were among Heidegger's most beloved: 'where, however, the danger is, grows / also the redeeming power' [*wo aber Gefahr ist wächst / das Rettende auch*] (Hölderlin, 1986, p. 193) Thinking cannot evade this danger. The kairos does not bring meaning to history, nor does it receive meaning from it. The threat of catastrophe cannot be neutralized by faith in redemption. Nevertheless, the danger far from excluding the redeeming, rather awakes a preparedness for it. To think this danger from the historical place in which a thinker is placed within a certain chronology, and with an awareness that human action within a kairos may lead to catastrophe, is to carry out the task and the responsibility of thinking.

Notes

Introduction

1 The present work takes as given Heidegger's thesis that 'eternity' is a derivative form of temporality. In this context the meaning of *αἰών*, which is translated as 'eternity' is undetermined. In the present work *αἰών* is for the most part understood in traditional terms as 'eternity', and as such is interpreted in terms of the circular recurrence of the eternal in contrast to the linearity of time. It is important to note here that the French philosopher Gilles Deleuze (cf. Deleuze 2005, pp. 72–4) interpreted the difference between *αἰών* and chronos as two different 'readings' of time. It is, however, beyond the scope of the present work to deal with Deleuze's account.

2 Augustine expressed this situation famously when he exclaimed: 'What is time? I know well enough what it is provided nobody asks me; but if I am asked what it is and try to explain, I am baffled.' Augustine, *Confessions* book XI, 17. This questionableness of time is one of the main motives for the phenomenological investigation of time. Cf. Husserl's reference to this passage in Husserl, 1964, p. 21.

3 Cf. Heidegger, 1984, p. 138; GA 26, p. 173, where Heidegger coins the neologism *Mannigfaltigung* (translated as multiplicity) to expresses this dispersion: 'This multiplicity does not occur because there are several objects, but conversely.'

4 Pöggeler, 1989, p. 288: 'I think however, that it is now about time, to stop writing *about* Heidegger. It would be more important to engage with the matter itself.'

Chapter 1

1 This quotation is from the student notes from a lecture course Heidegger held in the Summer Semester of 1934 which were part of the estate of Helen Weiss, a student of Heidegger's and the aunt of Ernst Tugendhat, and published in Spain without the permission of the Heidegger estate in 1991. The lecture course has subsequently been published as part of the *Gesamtausgabe* (1998), and an English translation of that edition was published in 2010. In those instances where the quoted passage is from the first, unofficial edition, I cite where appropriate the corresponding pages in the English and German *Gesamtausgabe* editions in that order.

2 'Historicity' is a relatively new philosophical term. It is employed by Hegel, but not as a terminus technicus. It becomes a technical term only through the influence of historicism, particularly in the work of Count Yorck and Dilthey. Dilthey stressed, rightly, that what was new in his method was the combination of being-in-the-world and history. Heidegger's use of the word is influenced by this. For a good account of this development of the concept cf. Renthe-Fink, 1964. 'Historicity' in the context of this work names a problem, namely that the human being – as a member of a community – is capable of relating to foreign worlds (either of the past or of the present). On the question of foreign worlds in Heidegger, cf. Brandner, 1994, pp. 130–4. This was not the way in which Dilthey understood historicity. For the latter, the historian, due to his historicity, was a part of that which he interpreted. Dilthey expressed this by saying that 'Life understands life'. Count Yorck, on the other hand, stressed the contradictions of history and the 'supernatural force' which could bring these contradictions into unity. Cf. Dilthey and Yorck, 1923, p. 61.

3 Cf. Heidegger, 2004, pp. 71–4; GA 60, pp. 102–5. For a comprehensive discussion of this, see Kisiel, 1993, pp. 151–91. Kairos is mentioned only once by Heidegger in the lecture courses around the time of the publication of *Being and Time*. In *The Basic Problems of Phenomenology*, Heidegger claims that Aristotle had not understood the 'now' radically enough. If he had done so, he would have had to come upon the phenomenon of the kairos, he says (cf. Heidegger, 1988, p. 288; GA 24, p. 409). The question remains as to why Heidegger only used the term kairos on this one occasion after 1921? This is a question I cannot definitively answer. However, Hans-Georg Gadamer ventured a suggestion in discussion with the author, suggesting that Heidegger did not want his position to be confused with that of the theologian Paul Tillich, who – in Heidegger's estimation – misused the concept in his writings. As an example of Tillich's use of kairos, see Tillich, 1963; cf. Ritter, 1971, columns 668–70.

4 First Letter to the Thessalonians 5, 2: 'For you yourselves know very well that the day of the Lord will come like a thief in the night."

5 Heidegger is expressing this when he says that 'the When is conditioned through the How of the self-comportment ' (Heidegger, 2004, p. 75; GA 60, p. 106)

6 The term 'kairology' is preferred here to 'kairotical', firstly to mark its correspondence to chronology, but also because I want to stress precisely the sense of the kairos, its meaning and its directionality.

7 I translate *Sein* with being and *Seiende* with entity or entities depending on the context. I avoid the confusion which arises with the other standard translations of these terms, whereby 'Being/being' suggests a difference between a substance and its modes, or 'being/beings' suggests a difference between singular and plural. The latter is the less misleading, but in the German both *Sein* and *Seiende* are singular terms, indeed are *singulare tantum* nouns.

8 Heidegger's analysis of Angst is well known, but it is noteworthy that, in *Being and Time*, he also mentions joy along with Angst, however without analyzing it in that work (Cf. BT, pp. 286, 317; H, pp. 310, 345).

9 The sense of kairos and kairological employed here assumes an interlacing of the Greek and Christian roots: the Greek in terms of the opportune, the Christian in terms of decision. The apocalyptical sense of the term is evident in some of Heidegger's later writings (especially in the *Contributions to Philosophy* [Heidegger, 1999; GA 65]) and such connotations are relevant to the question of the revolutionary, in particular the specifically Judeo-Christian sense of the messianic. Heidegger downplays these Christian themes in *Being and Time*, and especially in his writings in the 1930s, but it certainly seems to be the case that Heidegger glimpses in early Christianity a way of living temporality as kairos. Furthermore, what Heidegger sees in Paul is a manner to live possibility as possibility and not in terms of a 'possibility (the future) … fundamentally being tied to a present actuality' (Zangeheh, 2011, p. 545) (cf. Heidegger, 2004, p. 55; GA 60, p. 80).

10 Gadamer points to this event of the birth of Christ as marking a new sense of history: 'with this experience of a new covenant and with the Christian message of salvation history as history is discovered in a new sense' (Gadamer, 1993, p. 138).

11 Cf. Nietzsche, 1997, p. 64: The agent 'forgets most things, to do the one thing, he is unjust towards that which lies behind him, and he recognizes the rights only of that which is now to come into being and not other rights whatever.'

12 In this case, of course, the order is not endangered, but rather confirmed through this process. Furthermore the element of the surprising is, if not removed, at least formalized. For a similar observation, see Held, 1991a, pp. 310–11. Gadamer speaks of growing old and generational transition in terms of such discontinuity (cf. Gadamer, 1993, pp. 137–8).

13 Aristotle's question concerning the relation between time and motion lies at the roots of this double-sense. For Aristotle, time is neither identical with movement nor is it independent from it. Time belongs to movement. Time is that which one measures in movement. Without a measurer there can be no measuring, hence time depends on the intellect (*nous*); but time also belongs to motion. Cf. *Physics*, pp. 3, 11. This difference relates to, but is not the same as, that between subject and object.

14 Cf. Levi-Strauss' discussion of totemism in Levi-Strauss (1986). In his Kassel Lectures concerning Dilthey, Heidegger spoke of the unhistorical humanity: 'Primitive people live, and we ourselves lived a long time, without history.' (Heidegger, 2007b, p. 243; Heidegger, 1992–3, p. 145) If human beings have become historical, that does not make them any less historical. As such the recourse to a non-historical temporality is cut off. Furthermore, if the human has become historical and is essentially historical, what does that say regarding the

so-called primitive humanity, which (perhaps) is not (yet?) historical. On this issue cf. Carr, 1986, pp. 179–85.

15 This problem has been discussed very fruitfully by Hans Ruin (Ruin, 1994, pp. 145–75). Ruin's analysis coincides in large measure with that presented here, as I will indicate below.

16 *Wiederholung*, etymologically understood, means 'to fetch', 'to seek' or 'to summon' (*holen*) 'again' (*wieder*). This is a repeating, etymologically, *re-petere* – to go towards, to seek, to demand, to attack – or a retrieval, or indeed a reiteration, in which the past which is fetched again is so in its possibilities. Heidegger is here working through the account of repetition in Kierkegaard and recurrence in Nietzsche. Cf. Kierkegaard (1983) and Nietzsche (2001).

17 This first statement is confirmed by Heidegger: 'We must go back and free the ontological structures of Dasein already gained with regard to their temporal meaning. Eveydayness reveals itself as a mode of temporality. But by the repetition of our preparatory fundamental analysis of Dasein, the phenomenon of temporality itself at the same time become more transparent.' BT, p. 210; H, p. 234. That the first part of *Being and Time* represents a repetition of the history of ontology will be shown in this, and especially the next, chapter. On the concept of repetition in Heidegger, cf. Figal, 1988, pp. 31–40

18 On the problem of regressive (or progressive) method, which lies at the basis of this, cf. Heidegger, 1988, p. 280; GA 24, p. 397. With respect to time as the horizon of the question of being, cf. BT, p. 398; H, p. 437.

19 This defines phenomenology in Heidegger's sense. As he tells us, phenomenology's task is 'to let what shows itself be seen from itself, just as it shows itself from itself' (BT, p. 30; H, p. 34). This involves a movement from the ordering of entities, as they are ordered in Dasein's everyday experience, to the principle of order itself, which is for the most part hidden. What phenomenology seeks to reveal is that which 'does not show itself initially and for the most part something that is concealed, in contrast to what initially and for the most past does show itself. But at the same time it is something that essentially belongs to what initially and for the most part shows itself, indeed in such a way that it constitutes its meaning and ground.' (ibid., p. 31; H, p. 35). The 'meaning and ground' here is a principle of order, which, however, cannot be grounded, but rather is always presupposed. Heidegger's question concerning being is nothing other than a question concerning that necessary, but hidden, presupposition of all ordering.

20 The importance of the moment of vision – *Augenblick* – in Heidegger's thought (both early and late) is not to be underestimated. According to Pöggeler, '*Being and Time* can be understood as the first attempt of Heidegger's to work out a philosophy of the moment of vision.' (Pöggeler, 1994, p. 143). The thought of the

'other beginning' in the later work from the *Contributions to Philosophy* (1936–8) is premised on an understanding of time on the basis of the moment of vision.

21 This idea goes back at least to Aristotle and can be said to be the basic intuition of all substance ontologies. For an example from modern philosophy see Kant's first analogy of experience, *Critique of Pure Reason,* A 182/B224–A189/B232.

22 Heidegger, although recognizing his debt to Kierkegaard, when introducing the moment of vision in *Being and Time* unfairly states that he 'gets stuck in the vulgar concept of time' (BT, p. 412n. 338, n. 1), while in fact Kierkegaard's account of the moment is premised on a break with the substantialist assumptions of such a concept. Later in the *Fundamental Concepts of Metaphysics*, Heidegger is more generous, linking the thought of the moment to that of possibility in Kierkegaard (cf. Heidegger, 1995, pp. 150–1; GA 29/30, pp. 225–6).

23 This is a question to which Otto Pöggeler gave two distinct answers. In his *Martin Heidegger's Path of Thinking* (cf. Pöggeler, 1989, p. 170), he spoke of the moment of vision as empty. In a later essay, 'Destruktion und Augenblick', he speaks of it as authentic time (cf. Pöggeler, 1969, p. 20).

24 Heidegger does not speak here of the quiet reflective present in which – for a moment – the having-been and the future fall away. Heidegger's concern is rather with the present of action (cf. BT, p. 300; H, p. 326).

25 Cf. Kafka, 'Er', in Kafka, 1937, p. 300. For a discussion of this passage in Kafka, cf. Arendt, 1968, pp. 9–13. Cf. also Levinas, 1969, p. 69: 'The present is produced in this struggle against the past'.

26 Cf. BT, p. 344; H, p. 375: 'The specific movement in which Dasein is *stretched along and stretches itself along*, we call the *occurrence* of Dasein' [translation modified]

27 On the moment in Plato, see Beierwaltes (1969/70) and Link (1984). As far as I am aware, Heidegger discusses this phenomenon twice in his later work, in the *Parmenides* lecture course, and in the course *The Principle of Reason*. For a good analysis of these two brief discussions, see Ruin, 1994, pp. 205–7.

28 Although Heidegger read Simmel and was appreciative of his work, I am making no claim here as to the influence of Simmel on Heidegger's account of authentic temporality. Simmel's account is rather one which is illustrative of the question as to the temporal withdrawal from the everyday and the relation of that withdrawal to inauthentic temporality.

29 Eugen Fink made a comparison between the judgement concerning the dream in the sober light of morning and that of metaphysical philosophy concerning play. Cf. Fink, 1960, p. 137f.

30 Cf. also Heidegger, 1988, p. 276; GA 24, p. 391: Dasein 'is these possibilities' itself.

31 Although Bergson criticizes the concept possibility as such (cf. Bergson, 1968, pp. 100–1), the above analysis does not run counter to the tendency of his

thought. He too wishes to maintain a novelty of the future. His critique of Aristotle is implicitly a critique of the Greek concept of *poiesis*, which will be made thematic in Chapter 2.

32 This may seem a violent interpretation of Nietzsche. However, it cannot be ignored that, when Nietzsche speaks of the danger of history, he is referring to the danger of indifference (which lies at the basis of scientific disinterestedness). Monumental history will always 'have to deal in approximations and generalizations, in making what is dissimilar look similar', (Nietzsche, 1997, p. 70); antiquarian history 'hinders any firm resolution to attempt something new', (ibid., p. 75). Action out of forgetting forgets the past without destroying it; it has the plastic power to 'develop out of oneself in one's own way, to transform and incorporate into oneself what is past and forgotten … to recreate broken moulds' (ibid., p. 62).

33 In this chapter, I will use the word 'practice' without any reference to the difference between praxis and poiesis or between phrōnesis and technē. Under 'practice', I mean all relations of knowledge of the world, in the sense of a world that arises and changes out of and in human action, i.e. the world of human possibility. In the next chapter, I will deal with the specific difference between praxis and poiesis in the context of the question of eternity and Heidegger's claim to the priority of the future in temporality.

34 On this theme, see Wörner, 2011, pp. 67–9, who rightly points out that a claim I made in the original German version of this work that Aristotle has no account of kairological temporality cannot be sustained without qualification.

35 Cf. BT, p. 498n. viii; H, p. 385n. 1. For a representative text from Dilthey, cf. Dilthey (1957), pp. 36–41. As Barash points out, however, unlike Dilthey, in his account of generation Heidegger is referring not to an 'interrelation of appearances' but to ontological preconditions (cf. Barash, 1988, p. 171) On generation in Dilthey, cf. O'Byrne, 2010, pp. 46–69.

36 This is to make no claim regarding the disputed question of the 'right to life of the unborn', or indeed as to when human life begins. The point is merely that the entity which is in the womb prior to birth and the born child are the same.

37 On the structure of significance (*Bedeutsamkeit*) of public time, cf. BT, p. 380; H, p. 414.

38 Cf. Held, 1991a, p. 314: 'This own (the homeworld) can only be 'experienced' generatively, but cannot be 'appropriated' from outside.'

39 Cf. BT, p. 311; H, p. 339: 'We call authentic having-*been* repetition.' (translation modified).

40 Cf. ibid., pp. 358, 391: 'Authentic historicity understands history as the 'recurrence (*Widerkehr*)' of what is possible and knows that a possibility will only recur if existence is open to it fatefully, in the moment of vision (*Augenblick*), in resolute repetition.' (translation modified)

41 This account has been misread quite frequently as a mere submission to the past by many commentators. Karsten Harries, for example, reads this as a decision without any criteria and one which ends up as a subordination of the individual to the common destiny (Harries, 1978, pp. 650, 651). To see what is at issue here as a matter of subordination is to read *Being and Time* against the grain: at issue rather is the taking up of the past as possibility in the free space opened up by repetition and rejoinder. With respect to the choice of a hero in this respect, see Fynsk, 1993, p. 47. On the relation to heritage generally, see Young, 1997, pp. 71, 83–4.

42 BT, pp. 356–7; H, p. 390: 'Our lostness in the one and in the world-historical reveal[s] itself as a flight from death.' (translation modified).

43 Cf., BT, pp. 355–6; H, pp. 389–90. Heidegger explains Dasein losing itself in the dispersal with respect to the flight from death, which can only be overcome through resoluteness. On the chronological level, however, this does not occur. In the everyday, all Dasein can do is (re-)establish a unity without being aware of its lostness. Only in the kairos does Dasein face its abysmal lostness. Cf. ibid., p. 311; H, p. 338: 'In resoluteness, the present is not only brought back from dispersal in the objects of its closest concern, but held in the future and the having-been'. (translation modified).

44 This is not so in English or in French. In German, the difference is marked by the definite article: *der/die Erbe* = heir, *das Erbe* = heritage.

45 Dasein is not only being-with, but is authentically historical only in the same world with other Dasein, in a people, a community, a generation. But Daein is historical because it is fated. And it is fated only because it lives with others in the same world. If that is the case, then Dasein is ontologically and existentially historical only because it is factically together with other Dasein – its generation. This raises questions as to the transcendental structure of *Being and Time*. This is a topic to which I will return below in Section III.

46 BT, pp. 344, 347–8, 349, 352, 357; H, pp. 375–6, 379, 382, 385, 391

47 On the question of discourse in the moment, cf. Wohlfart, 1982, pp. 136–60 and Pöggeler, 1989, p. 170. The problem of the linguisticality of the moment cannot detain us here. In contrast to Wohlfart, my concern is with action in the moment, not with the related aesthetic questions. For an account of the Augenblick from a literary perspective, cf. Bohrer, 1981.

48 Implicit here is a distinction between practice and *poiesis*, which I will return to in Chapter 2.

49 Cf. BT, p. 265; H, p. 288; Heidegger, 1985, p. 319; GA 20, p. 441; Heidegger, 2007b, p. 266; KV, p. 168.

50 Cf. BT, p. 311; H, p. 338: 'the resolute rapture of Dasein, which is yet held in resoluteness, in what is encountered as possibilities and circumstances that are of concern in the situation'. (translation modified).

51 '*Befindlichkeit*' is one of the most difficult words in *Being and Time* to translate, as is evident in the unsatisfactory nature of both Macquarrie and Robinson's and Stambaugh's efforts. 'State of mind' (Macquarrie/Robinson), although it does capture that the term refers to a state of being in which Dasein finds itself, is misleading with its mentalist suggestions. '*Attunenment*' (Stambaugh) is a strange choice as it is a better translation of '*Stimmung*' ('mood' in both translations). *Befindlichkeit* comes from *sich befinden*, which refers to a state of well-being or how someone is feeling. I translate with 'affective state'.

52 Cf. Ruin, 1994, p. 165: 'To speak of Dasein's possibilities as heritage suggests that the temporalization of understanding cannot be thought apart from a certain belonging of Dasein to its past, an original opening to the other, which is also an opening to language.'

53 The present financial crisis illustrates this phenomenon well.

54 Cf. BT, p. 161; H, p. 172: 'Circumspection gives all our teaching and performing its route of procedure, the means of doing something, the right opportunity, the proper moment.' In his critique of the kairological reading of Augenblick, Hakhamanesh Zangeneh takes the author, amongst others, to task for missing the inauthentic nature of the opportune moment, citing this passage (cf. Zageneh, 2011, p. 552n. 30). The point, though, is that there the opportune is on the ontic level only because of the order of entities which arises suddenly, in the moment of vision, for a Dasein which understands its being in terms of possibility as possibility. The opportune moment here is not one of the 'in order to (*wozu*)', but rather of ontological understanding and – how odd the term may sound – philosophical action, that is action which responds to the disclosure of being as possibility in disclosing the contingency of the actual worldly order.

55 On the political and ethical significance of philosophy in its withdrawal from the political and the ethical, see Ó Murchadha, 2012.

Chapter 2

1 For the remainder of this book, I will write 'praxis' and 'poiesis' (and also *technē*) in Latin script and will use them as English words.

2 My interpretation of this difference in *Being and Time* is above all indebted to a justly famous article by Robert Bernasconi: 'The Fate of the Distinction between Praxis and Poiesis' (Bernasconi, 1994). Of great significance also are Volpi, 1988, McNeill, 1999, pp. 93–136 and Taminiaux, 1991, pp. 147–81.

3 I am here following Franco Volpi's thesis that *Being and Time* is a repetition at an ontological level of Aristotle's *Nicomachean Ethics*. Cf. Volpi, 1988.

4 Stambaugh translates '*Bruch*' with 'breach' and '*die Umsicht stößt ins Leere*' with

'circumspection comes up with emptiness'. Macquarrie and Robinson translate similarly. The German though has a more violent and disruptive connotation. '*Bruch*' has the meaning of disruption, of a sudden break in which something breaks off and does so because of some injury endured; '*stoßen*' has the meaning of bumping into, but here to bump into that which blinds it, hence to crash.

5 Gethmann (1989) has already criticized Prauss on this point: 'The primary dealing has a cognitive moment, which however is not a condition (that is logically prior to the dealings), but rather a characteristic (logically contemporary)', pp. 171–2n. 12.

6 Cf. BT, p. 65; H, p. 69: 'What is peculiar to what is initially at hand [*Zuhandenen*] is that it withdraws, so to speak, in the character of handiness in order to be really handy.'

7 I use 'environment' to translate '*Umwelt*' (following Maquarrie and Robinson) not alone because it is the normal translation of this term, but also because it expresses well the difference from the world as the horizon of appearance, which is central to this discussion. Stambaugh's rendition of 'surrounding world' although it captures the etymological meaning of the German (*um* – surrounding, *Welt* – world) obscures the issue.

8 I have in this instance preferred the Macquarrie/Robinson translation, cf. Heidegger, 1962, pp. 96–7.

9 Cf. Heidegger, 1988, pp. 106–12; GA 24, pp. 149–58 and 'Plato's Doctrine of Truth', in Heidegger, 1998, pp. 155–82; Wm, pp. 201–36.

10 According to Heidegger, the Greeks failed to see the 'abysmal problematic', that the time is being understood on the basis of time, cf. Heidegger, 2010, p. 163; GA 21, p. 193. That time, which should be grounded in being, itself is the ground of being, this brings us before the abyss (*Ab-grund*). Heidegger's response to this abyss is '*to understand time from out of time*' (Heidegger, 2007a, p. 200; BZ, p. 6)

11 Cf. Heidegger's interpretation of the 'what-being' of the entity as *τό τι εν ειναι* in Heidegger, 1990, p. 164; KPM, p. 240.

12 See page 64 above.

13 In this sense, the useful thing is a symbol in the original sense of the word, i.e. 'a recognition sign … for example a broken coin, the two halves of which would be divided between friends who would be a distance apart', Fink (1960), p. 118. On the concept of symbol as an approach to the understanding of the world, cf. Fink (1960), pp. 118–21.

14 This whole way of reading Heidegger – in terms of praxis and poiesis – has been attacked in fundamental terms by Graham Harman (cf. Harman, 2002 and 2007). It is impossible to do justice to the richness and radicality of Harman's interpretation of Heidegger here. Nevertheless, the main argument can be addressed. For Harman, handiness is not specific to equipment, but characterizes

the withdrawal of all things – things in their withdrawal he terms 'tool-being' (Harman, 2002, p. 24: 'tool-being … is irreducible to anything that could be seen.') – which is the reverse side of objective presence – in his terms 'broken tool' (Harman, 2002, p. 46: 'Heidegger's thought starts from the universal dualism between tool and broken tool'). Heidegger's insight is into this relation of tool-being and broken tool, which is then repeated monotonously through his philosophy. A consequence of this is that the issue here is not to do with human practice, is not to do with any distinction between doing and making, but rather is a universal ontological claim. I also understand handiness not as the singling out a particular class of entity, but as revealing a way of being of all entities. Nevertheless, Heidegger continually tells us that Dasein is neither handy nor objectively present. This can be understood – as Harmann suggests (e.g. ibid., pp. 16, 18–19, 33–4) – as the 'excessive role of Dasein' in *Being and Time*, but might it not more profitably be seen as the claim that, for handiness to appear at all, it must appear within a horizon which allows it to be seen? What allows it to be seen is world. World can only appear for an entity which transcends entities towards being. Without such transcendence, all we have are things 'encountering' other things (cf. ibid., p. 30). Furthermore, Harman's interpretation of Heidegger works only at the price of excluding whole tracts of Heidegger's account (in *Being and Time* and elsewhere) from consideration. But it is a doubtful hermeneutical strategy which succeeds only in eliminating from consideration – or setting aside, as incidental to Heidegger's principal discovery – such themes as authenticity, death, mood, freedom, conscience, historicity and the transcendental structure of *Being and Time* itself. Furthermore, against Harman, I don't think there is an opposition here between history and phenomenology (cf. ibid., pp. 112–14): Heidegger's retrieval of Aristotle's difference of praxis and poiesis is not simply doing the history of philosophy, but is a destructuring of the history of philosophy, where the motivation for this destructuring is phenomenological: the uncovering of experience as fundamental to its philosophical articulation.

15 Cf. BT, p. 137; H, p. 146: 'Understanding can turn primarily to the disclosedness of the world; that is, Dasein can understand itself initially and for the most part in terms of the world. Or else understanding throws itself primarily into the for-the-sake-of-which, which means Dasein exists as itself.'

16 Heidegger hints at the logic of failure when he says 'All failure (*Versagen*) is in itself a saying (*Sagen*), that is a making manifest.' Heidegger, 1995, p. 140; GA 29/30 p. 211. The logic of reciprocal relation, however, calls into question Heidegger's own self-understanding of the ordering principle of *Being and Time*.

17 Cf. for example, http://news.harvard.edu/gazette/story/2010/11/partial-reversal-of-aging-achieved-in-mice (accessed 21st February 2012).

18 In the lecture 'What is Metaphysics?', Heidegger gives a long analysis of Angst

without thematizing death or mortality (cf. Heidegger, 1998, pp. 87–93; Wm, pp. 110–17).

19 Cf. on this Heidegger, 1985, p. 291; GA 20, p. 403: 'Angst is nothing other than the pure and simple experience of being in the sense of 'being-in-the-world. This experience can, though it does not have to – just as all possibilities of being come under a 'can' – assume a distinctive sense in *death*, or more precisely, in *dying*.' My question is what conditions the choice of this possibility. To answer this we must explore the basic experience of Dasein which Angst disrupts and ruptures.

20 Heidegger, on the contrary, always emphasizes that as long as it is, Dasein is always in the movement of projecting, (cf. BT, p. 167; H, p. 179).

21 On the circularity (the hermeneutical circle) of understanding, cf. ibid., pp. 142–4, 152–3. The circle is always provisional. It can so little encompass Dasein, that Dasein must first come into the circle. As possibility, Dasein has always already projected itself, but it can never completely fulfil this projection. Cf. ibid., pp., 136, 145: 'the project character of understanding means that understanding does not thematically grasp that upon which it projects, the possibilities themselves. Such a grasp precisely takes its character of possibility away from what is projected, it degrades it to the level of a given, intended content, whereas in projecting project throws possibility before itself as possibility'.

22 On death as originally actuality rather than possibility, cf. Müller-Lauter (1960), pp. 39–53. But see my later discussion of Müller-Lauter's thesis in Chapter 3.

23 Cf. Waldenfels, 1987, p. 12. Waldenfels speaks here of an asymmetry between this side and the far side of the threshold. This asymmetry is, in one sense, that which makes things possible, such that only as a living being can one talk about life and death. But when we forget this asymmetry and speak about death as if we could go beyond the threshold and besides that speak about life as if we stood beyond the threshold, then we stand in an illusory position.

24 Carr is right to interpret futuricity in this way with respect to being-towards-death, but as will be shown later, Heidegger's account of futuricity allows for another interpretation, cf. Carr, 1986, p. 343; Ó Murchadha, 1998.

25 Whether a person could be declared happy before her death is an issue which Aristotle already discusses, and with some ambivalence answers in the positive (cf. NE, 1101a10–1101a20). My concern here though is with a more everyday sense of happiness and success, which until death is always subject, as Hamlet tells us, to the 'slings and arrows of outrageous fortune'.

26 Cf. BT, p. 223; H, p. 240: 'In "ending", and in the totality thus constituted of Dasein, there is essentially no representation.'

27 Above all in his lecture course of 1929–30. cf. Heidegger, 1995, pp. 60–6; GA 29/30, pp. 91–9. See too Heidegger's use of this term in the 1930s, especially in Heidegger, 2002b, p. 151; GA 34, p. 209 and Heidegger, 1987, p. 179; EiM, p. 137.

28 We can see here again how the present can seem to disappear into the future and the having-been. Waking does, however, have a moment of resistance. It resists the past insofar as it calls for transformation; and to the extent to which it brings to light the simultaneity of away- and there-being (*Weg- and Da-sein*), there is in it a moment of hesitation before the leap into the future.

29 The old order can simply continue, but it must continually justify itself. This need for justification is a sign that the futuricity of Dasein is that of a historical entity. Only because the past cannot simply be transmitted into the future as an enduring and self-evident order is there the need to justify such transmission, to justify, namely, tradition. This is at the same time a need to change the order of the past in the 'today' – either incrementally or structurally.

30 In contrast, that is, to the tradition of philosophy from Plato, for which philosophical theory is a sight for the dying. Cf. *Phaedo*, 64a4–5.

31 Cf. Heidegger, 1988, p. 277; GA 24, p. 393. For this and what follows, cf. Volpi, 1988. It is significant that Heidegger characterizes the historicity of Dasein as understanding in the same context as he speaks of authentic action.

32 Cf. NE, 1140 b 6–7: 'production aims at an end other than itself; but this is impossible in the case of action, because the end [*telos*] is merely doing well.'

33 In his lecture course for the winter semester 1924–5, Heidegger shows that *βούλησθαι* is not the observation of anything present, but rather of something which needs to be sought, which is not yet present, but needs to be uncovered, cf. Heidegger, 2003, 102; GA 19, p. 149. In other words, praxis can only be understood in terms of futuricity for Heidegger and, if temporality is to be understood as a constitutive possibility of Dasein, then futuricity has to be understood otherwise than in teleological terms. This is so because what is at issue is not possibility in the sense of something actualizable, but possibility of the radically new.

34 Cf. Figal, 1988, p. 92: 'Dasein should not be understood, however, as goal or end in an Aristotelian sense, even if everything one does is done 'for the sake of' Dasein. If possibility is the final positive determination of Dasein, there cannot be such an actual completion.'

35 One might object that there are indeed times for the making of certain things dictated by time. One example of this is the making of baskets from reeds as practised in the west of Ireland. In this case, there is a certain time of year at which the reeds can be harvested and only at this time can the baskets be produced. This, I submit, is a limit case. It involves not so much a particular time as it does a season and is a case which lies between production and agriculture (in Arendtian terms, between making and labour). Agriculture works with the biological cycle of nature; in praxis, in the kairos, it is a matter not of acting in a cyclical time, but rather of transcending such cyclicality. I would like to thank Prof. Markus Wörner (NUI, Galway) for bringing this case to my attention.

36 The debate regarding the question of politics in *Being and Time*, and in Heidegger's thought before 1933 generally, has in recent years been marked by the attempt to read that work in the light of Heidegger's later engagement with the Nazis. I will address that debate in Chapter 5, and will refer to it only sporadically and in footnotes in this section. What does tend to get lost in this whole debate is the question of the possibility of a political world as such. It is that theme which I wish to explore, if somewhat indirectly in the next section. For a point of orientation, here, see Held, 2010.

37 On this issue, see also BT, pp. 110–11; H, pp. 117–18, where Heidegger speaks of the other in terms of the world of the handworker, as producer and supplier.

38 It would undoubtedly prove fruitful to bring the following analyses into dialogue with Derrida and, indeed, Carl Schmidt. I do not have the space to do so here (cf. Derrida, 2000 and Schmitt, 2006).

39 I cannot tackle here the problem of spatiality, but the impossibility of understanding such a temporalizing without reference to space is an indication that the priority of temporality in *Being and Time* is perhaps tied to the emphasis on the individualized Dasein.

40 However, he does name, but then only in passing, one example, namely, walking through a city centre on a Sunday afternoon (specifically when all the shops are closed); cf. Heidegger, 1995, p. 135; GA 29/30, p. 204.

41 Such responsibility cannot be understood, however, without reference to the delivering over (*Überantwortung*) of Dasein to its possibilities, which makes up thrownness.

42 Heidegger suggests this reference himself when he describes the '*Ansagen*' in '*Versagen*' as call, cf. Heidegger, 1995, p. 135; GA 29/30, p. 216. On the question of the compulsion to listen, cf. Held (1991), p. 36.

43 Only in two contexts does the concept of power or domination appear in *Being and Time*: in the analysis of conscience and in the analysis of historicity; cf. BT, pp. 254, 257, 286, 352; H, pp. 275, 278, 310, 384–5.

44 Heidegger is admittedly ambiguous on this in *Being and Time*, and there is an emphasis in that work which makes it hard to disagree with Jean Grondin when he notes: 'The reader of *Being and Time* finds a Dasein which remains the master of its own existence, that which signifies that Dasein is, in a certain sense, the author of its temporality.' (Grondin, 1987, p. 88) But this is neither the only way in which this relation can be understood in that work nor the most compelling.

45 As such, the ambiguous power of time is expressed most fully in Heidegger's understanding of conscience. Temporality and conscience are structured as relations to that which works through the self but is experienced as having a source transcending the self.

46 Cf. BT, p. 286; H, p, 310: 'Together with the sober Angst that brings us before our individualized potentiality-of-being, goes the unshakable joy in this possibility.'

47 Cf. note 16 above.

Chapter 3

1 Causality is being understood here in its modern sense of *causa efficiens*, because Heidegger, in the wake of *Being and Time*, engages is a sustained critique of the modern account of cause, particularly in his discussions of Kant. On the development of the concept of freedom in modernity, see Zakira (1994).

2 For Descartes cf. Descartes, 1984, p. 31. In opposition to Descartes, Heidegger understood infinity negatively. C.f. BT, p. 304; H, pp. 330–1. For Heidegger's interpretation of Descartes on the question of God, cf. ibid., pp. 86–8; pp. 92–5.

3 The exchangeability of 'accident' and 'contingency' is not Aristotelian. Rüdiger Bubner, in an Aristotelian vein, expresses the distinction between the two in this way: '"contingency" is the arising of arbitrary alternatives without reason; "accident" is that which we cannot predict, which appears to occur with reason, but which could have arisen from purposeful action.' (Bubner, 1984, pp. 35–8) This distinction is not being employed here because what I wish to show is the groundlessness of action in the kairos.

4 Cf. also Schwan, 1989, p, 249. We can find this experience of becoming strange in Heidegger's own mode of interpreting. He attempts in an ever renewed manner to interpret texts such that their strangeness is presented. In this way, every genuine interpretation produced a transformation.

5 This is what Heidegger has in view when he speaks of the radicalization of the repeated through repetition, cf. Heidegger, 1988, p. 316; GA 24, p. 449. Cf. also Heidegger, 1987, p. 39; EiM, pp. 29–30: 'we do not repeat a beginning by reducing it to something past and not known … the beginning must be begun again, more radically, with all the strangeness, darkness, insecurity that attend a true beginning.'

6 This ambivalence in Dasein's relation to the past lies at the basis of Heidegger's project of the destructuring of the Western metaphysical tradition. As he puts it: 'Negatively, the destructuring is not even related to the past: its critique concerns "today" and the dominant way we treat the history of ontology … destructuring does not wish to bury the past in nullity; it has a *positive* intent.' (BT, p. 20; H, pp. 22–3). The critique of the 'today', however, concerns the past insofar as the latter still has power in the present. The positive intent is to come face to face with the past in its own right precisely through a destructuring which reveals its strangeness and discontinuity with the present.

7 Heidegger sees the greatest possibility of philosophy in this between-space. In discussing his last conversation with Max Scheler, he says that they were agreed about the 'essentials', namely, that 'the moment is here, now when the official philosophical situation is hopeless, to risk again the step into authentic metaphysics, that is, to develop a metaphysics from the ground up.' (Heidegger, 1984, p. 132; GA 26, p. 165) Precisely when the situation seems to be at its worst, hence when the old order is at its most questionable, the possibility for philosophy is at its greatest.

8 As such, the will in *Being and Time* is, as Klaus Held correctly states, 'not to be understood as a spontaneous active ability to dispose of things, but rather as the capacity to perceive that which is said to Dasein in a fundamental mood' (Held, 1991, p. 41.

9 Heidegger hints at such a phenomenon when he distinguishes between the time of peeks and of valleys. 'The time of the peak is long, because upon the peaks there rules an incessant waiting for and patiently waiting on [*Warten und Harren*] the event [*Ereignis*], without boredom and without pastime [*keine Langeweile und kein Kurzweil*]'. (Heidegger; 1989, p. 56). I will return to this relation of waiting to intensity.

10 It is to misinterpret Heidegger's account of authenticity to understand enticement and seduction as necessarily inauthentic. As with Augustine, for Heidegger the problem of falleness is not that the human being lets himself be enticed; it is that he lets himself be enticed by the wrong 'thing'. For Heidegger, authentic enticement is that which draws Dasein to transcend entities towards the meaning of being.

11 Cf. 'On the Essence of Grounds' in Heidegger, 1998, p. 123; Wm, p. 157: 'only if entities having the character of being-in-the-world irrupt into being, is there the possibility of entities manifesting themselves.'

12 Cf. Heidegger, 1991a, p. 7, where Heidegger represents interruption and liberation as two sides of transcendence.

13 This would only be the case if history was governed by causality, where 'if this entity exists as conditioned, then what conditions also exist, i.e. the complete series of conditions and the unconditioned itself must certainly exist.' (Heidegger, 2002a, p. 158; GA 31, p. 231)

14 Cf. BT, p. 135; H, p. 144 'And since understanding is affected and the affective state is existentially delivered over to throwness, Dasein has always already gone astray and failed to recognize itself' [translation modified].

15 Cf. ibid., p. 197; p. 213: 'The phenomenon of truth has always been one of the themes of our earlier analyses, although not explicitly under this name.'

16 C.f. ibid., p. 216; p. 234: 'by thus repeating our preparatory fundamental analysis of Dasein, the phenomenon of temporality itself will at the same time becomes

more transparent.' Tugendhat, on the contrary – and in my view mistakenly – argues that §44 is an appendix to Section 1 and is not integral to its analyses. (cf. Tugendhat, 1970, p, 259).

17 The fact that Dasein does not seek first truth, but necessarily encounters it, shows that following the 'practical' interpretation of the temporality of Dasein attempted here, the paradox of the Socratic situation (already discussed in Chapter 1) can not so much be solved as avoided. Truth is found, although not sought, because Dasein can only deal with entities insofar as it has the character of possibility. If this being possible is disclosed, it is not that Dasein is referred to a truth which was always already there, rather Dasein is pointed towards the groundlessness of truth; truth without signs to guide the way (cf. BT, pp. 288 and 145; H, pp. 312 and 154–5).

18 Cf. ibid., pp. 301–2; p. 328 [translation modified]: 'Resolute, Dasein has brought itself back out of fallenness in order to be all the more authentically "there" for the disclosed situation in the "moment of vision (*Augenblick*)"'. A little earlier Heidegger describes resolute being as the 'letting [of] *what presences* in the surrounding world be encountered in action.' ([ibid., p. 300; p. 326 [emphasis in original]).

19 Cf. ibid., p. 206; p. 225 (emphasis in original): '*to the extent that in this discoveredness, as a* discoveredness of ..., *a relation to things objectively present persists, discoveredness (truth) in its turn becomes an objectively present relation between objectively present things* (*intellectus et res*).'

20 Cf. Figal, 1988, p. 380: '"correctness" is not only and not primarily to be understood as a possible characteristic of propositions, but rather names the orientation towards a "direction setting" *model of actuality*, which one can correspond to in thinking as in conduct, in order to find a constancy concerning what and how one is'. Cf. also Plato's Doctrine of truth, in Heidegger, 1998, pp. 177; Wm, pp. 228–9: 'As unhiddenness, truth is still a fundamental trait of entities themselves. But as the correctness of the "gaze", it becomes a characteristic of human comportment towards entities.'

21 This does not mean that every propositional truth is true only in a certain now, but only that the verification of the truth of a judgement is always bound up with a now.

22 The words 'understood as the correctness of the proposition' is only contained in the edition of *Wegmarken* published in the *Gesamtausgabe* (Volume 9). In the earlier editions, the sentence simply read 'The essence of truth is freedom'. For a discussion of this variation, cf. Bernasconi, 1985, p. 79. In any case, the later addition corresponds with the context of the sentence.

23 This is the claim of both Tugendhat and Schwan. Cf. Tugendhat, 1970, p, 378, where he speaks of a 'turn from freedom to truth' and Schwan, 1989, p. 233: 'The fundamental topos of freedom is replaced by that of truth'.

24 Heidegger hints at this background when he refers to mood, cf. Heidegger, 1998, p. 147; Wm, p. 189.

25 The radical turn here from the position taken in *Being and Time* is evident. There Heidegger states the opposite: 'Only because Dasein exists as constituted by disclosedness (that is by understanding) can something like being be understood, only so is an understanding of being possible at all' (BT, p. 211; H, p. 230)

26 Heidegger poses this problem in a marginal note to his copy of the lecture course 'On the Essence of Human Freedom': 'The problem of the authority [*Instanz*] of the proof of projection. To the extent to which it has occurred and has done so as a whole the proof or refutation lies on the side of the dialogue partners [*Mitredenden*] not of the one who projects. So truth amounts to irrefutability. Absolutely not! What then?' (Heidegger, GA 31, 181, n. 8; note not reproduced in translation)

27 This problem is understood in Being and Time as concerning the relation between the existentiell and existential spheres (cf. BT, p. 288; H, pp. 312–13), because Heidegger had not yet at that stage posed the question how this difference, and with it the ontological difference, could arise at all. This difference is assumed, but not questioned in that work. Only when Dasein as that entity which is concerned for its own being came to be understood as a historical occurrence, could the question of the ontological difference itself be posed. In the absence of such a conception the 'ground' of this difference must be presumed to lie in the structure of Dasein's being. As such it remained inaccessible, because presupposed in every thematization. I will return to this question in Chapter 4.

28 Heidegger, 1998, p. 152; Wm, p. 196: 'In the same historical moment of vision in which the beginnings [*Anfang*] of philosophy fulfill themselves [*erfüllt*], the marked domination of common sense (sophistry) also starts [*beginnt*]' [translation modified]

29 This difference between liberation and self-binding relates to the traditional question of negative and positive freedom. Cf. Heidegger, 2002b; pp. 43–4; GA34, pp. 58f.

30 Cf. Heidegger, 1998, p. 150; Wm, p. 194: 'Errancy is the free space for that turning in which in-sistent ex-sistence adroitly forgets and mistakes itself constantly anew.'

31 Cf. ibid., p. 148; p, 192: 'And if the human being sets out to extend, change, newly assimilate, or secure the openedness of the entities pertaining to the various domains of his activity and interest, then he still takes his directives from the sphere of readily available intentions and needs.'

32 On curiosity cf. BT, p. 161; H, p. 172: Curiosity 'seeks novelty only to leap from it again to another novelty.'

33 Cf. Dahlstrom (2001), p. 278: 'disclosedness … makes it possible for entities to be uncovered and concealed and, thus, for propositions to be true or false,

but neither as a cause that produces such phenomena nor as a condition that somehow obtains apart from them. Instead dislosedness makes perceptual and propositional truths and illusions possible by co-constituting the uncovering, obscuring, or concealing.' This co-constitution means, on the one hand, that there is disclosure only insofar as there is discovery; and, on the other hand, not only is disclosure one condition of discovery, it is also part of the structure of discovery. The disclosure of possibility in the moment is only thinkable if Dasein stands in a chronological relation with a real entity which discloses itself. Furthermore, the dealings with an actual entity is not only made possible through the possibilities of this entity and those of Dasein, but is rather essentially moulded by them. Dahlstrom is putting forward a transcendental interpretation of Heidegger's account of truth, which he opposes to a pragmatic interpretation. As already pointed out, the standard pragmatic interpretation of Heidegger is based in a undifferentiated concept of praxis. If, however, the transcendental level is interpreted as the transcendental structure of action itself, then such an interpretation is not in contradiction with the one put forward here.

34 If there was not at least a vague recognition of the difference between the order of entities and the entities themselves it would be impossible to understand why the prisoners in Plato's allegory of the cave would be able to pose the question of truth (*alētheia*) in the first place. Only in the sense that there is something beyond the shadows – something hidden – allows the prisoners to understand the shadows as the unconcealed. Cf. *Republic*, VII, 515c1.

35 Heidegger says that untruth is the essence of truth. This does not mean that each propositional sentence is both true and false. Rather, Heidegger means that every dealing with entities is both revealing and concealing. In our example, if one says 'the window is closed', the window in its functionality must be first disclosed, but the order of possibility in which it stands, which in every understanding is assumed, remains hidden.

36 Cf. Heidegger, 1995, p. 147; GA 29/30, p. 221: 'Bound by time, Dasein cannot find its way to those entities that *announce* themselves *in the telling refusal of themselves* [*als das sich versagende*] as a whole precisely within this horizon of binding time.'

37 BT, p. 286; H, p. 310: 'Together with sober Angst … goes … unshakeable joy.'

38 Kierkegaard, indeed, speaks of the 'waltz of the moment of vision', in Kierkegaard, 1980, p. 108.

39 Certainly the time of using equipment can have its own rhythm – the rhythm of hammering, for example – but this is not constitutive of the directional relation of the in-order-to structure of such action.

40 Cf. on this Nietzsche, 2001, p. 84: 'Rhythm was supposed to make a human request impress a god more deeply'; ibid: 'rhythm is a compulsion … not only the stride of the feet, but also the soul itself gives in to the beat.' According to

Nietzsche, the Greeks believed that hexameter had been invented in Delphi, ibid., p. 85.

41 That this sentence is problematic in a number of respects will be discussed in the conclusion to this work.

42 Jaspers, 1989, p. 118: 'In enthusiastic striving ... the view is full of longing and drive and has wandered away from the present situation. Dissatisfied, without rest, but at the same time fulfilled and enraptured by that which it gazes upon the soul is moved and experiences relapses and uplift.'

43 Cf. ibid., p. 120: 'Who consciously wagers his life, experiences a unique freedom.'

44 Cf. Haar, 1988, pp. 266–7. It is important to state here that, in the context of his interpretation of Hölderlin, Heidegger brings mood and language into harmony with one another in poetry. This was something he had not achieved in *Being and Time* although one does find the following sentence in that work: 'The communication of the existential possibilities of attunement, that is, the disclosing of existence, can become the true aim of 'poetic' speech.' (BT, p. 152; H, p. 162.)

45 Cf. BT, p. 127; H, p. 134: 'The pure 'that it is' shows itself, the whence and whither remain obscure.'

46 The inner connection of truth and preservation is indicated in German by a linguistic connection: *Wahr*heit = truth, be*wahren* = to preserve.

Chapter 4

1 C.f. Dreyfus, 1984 and Zimmermann, 1990, pp. 166–90. This relates directly to the political question in the next chapter. The claim is not that Heidegger's concerns about technology were absent from his thought in general, but rather that they were not well developed philosophically in his thought in the early or indeed mid-1930's.

2 At this stage this remains a claim, which will be defended in the third part of this chapter. My thesis is: philosophy founds nothing because it always moves in the direction of the lack of grounds, the abyss. The question which then arises is whether philosophy can itself be understood as a work. This thesis is both sustained and challenged by Heidegger's thought in the 1930's, in particular in his account of the relation of philosophy and poetry. As Jennifer Gosetti-Ferencei has argued, there is here what appears as 'an ambivalence, and sometimes as a contradiction' in Heidegger's account of poetic language between a critique of violence on the one hand, and a belligerent and violent founding will on the other. (cf. Gosetti-Ferencei, 2004, pp. 6–7) This ambivalence is at the heart of Heidegger's account of the work, of poetry, art, politics and the relation of philosophy and thought to them in this period.

3 The piece of equipment can nonetheless find a place again in today's world. In ecological agriculture, for example, one finds old pieces of equipment such as the scythe, which had not been used for some time. In this case, a historical piece of equipment has been given a new, renewed significance in a new world.

4 Heidegger explicitly rejected this solution: 'How are these useful things historical when they are, after all, *not yet* past. Only because they became an *object* of historiographical interest, of the cultivation of antiquity and national lore? But such useful things can only, after all, be historiographical objects because they are somehow in themselves historical.' (BT, p. 348; H, p. 380).

5 On the difference between animal and human Dasein in Heidegger see Heidegger, 1995, pp. 186–267; GA 29/30, pp. 274–388. Cf. also Derrida, 2008, pp. 58–70 and Lawlor, 2007, pp. 46–60.

6 It might be objected that I am implicitly accepting that the piece of equipment makes a claim on Dasein by distinguishing it as peculiar and strange. But as will become clear it is not the tool-being of the tool, but the work through which we experience the latter which makes a claim on us: we experience 'the equipmentality [of equipment] properly only through the work' (Heidegger, 2002, p. 43; Hw, p. 56)

7 The being of a people is always unique, because it is historical. The historicity of a people depends on the uniqueness of a happening of being at a particular place, at which being is established (*gestiftet*). Cf. 1989, p. 121. The concept of people (*Volk*) has of course become a controversial point of contention in the debate regarding the relation of *Being and Time* and Heidegger's Nazi engagement. This is a topic to which I will turn in Chapter 5.

8 Heidegger refers approvingly here to Count Paul von Yorck. It is the case that Yorck praises Dilthey's concept of type as a 'historical category' (Dilthey and Yorck, 1923, pp. 190–3). But he criticizes Dilthey in a way which corresponds to the intention of Heidegger's critique of the uprooting of the historicity of Dasein, namely the 'procedure of comparison' (ibid., p. 193) and its aesthetic presuppositions.

9 Cf. Heidegger, 1984, p. 158; GA 26, p. 201: 'The art of existing is not the self-reflection that hunts around uninvolved, rummaging about for motives and complexes by which to obtain reassurance and a dispensation from action. It is rather only the clarity of action itself, a hunting for real possibilities.'

10 Cf. ibid., p. 156; p. 198: 'The finitude of philosophy consists not in the fact that it comes up against limits and cannot proceed further. It rather consists in this: the singleness and simplicity of its central problematic, philosophy conceals a richness that again and again demands a renewed awakening.' The simplicity of this central problematic of philosophy cannot be generalized. Philosophy must always be awoken anew because it is not concerned with general concepts, but with the singularity of fate (cf. BT, p. 351; H, p. 384), which only happens uniquely.

11 In his years at Marburg, Heidegger had of course worked intensively on theological issues. The relation of philosophy to this theology was very much a foundational one for him, as is evident from his lecture 'Phenomenology and Theology' (cf. Heidegger, 1998, pp. 39–62; Wm, pp. 45–78). In the 1930s, his interest turns to god and gods as they emerge in the work of Hölderlin and Nietzsche. A major impetus for this renewed interest, and a moulding influence on his understanding of the gods, was the work of Walter F. Otto, especially *Die Götter Griechenlands*, which was first published in 1929.

12 Cf. Heidegger, 1989, p. 188: 'Dionysius [the most excellent demi-god] brings the trace of the flown gods down to the godless.'

13 Cf. ibid., p. 172: '"Fate" is the name of the be-ing [*Seyn*] of the demi-gods.'

14 It was Marx who first saw this problem in relation to the mask. In his essay 'The Eighteenth Brumaire of Louis Napoleon', he says that in a revolutionary situation when the prose of the everyday no longer suffices, the revolutionary falls back on poetry. He wears the masks of past heroes, because he falls back in horror before the new. According to Marx, only when human beings no longer rely on masks, but rather gain the power to act in the moment in full consciousness that their actions are not the actualization of past ideals, does the possibility of a socialist revolution open up. (cf. Marx, 1977, pp. 398–400) The mask, however, which has no face but only hides a further mask, indicates the power of transformation itself, in which Dasein must immerse itself during a revolutionary time. In Heidegger's terms, Marx was indeed right that the capacity to act in a revolutionary situation is the ability to respond to the claim of the moment, but this capacity is that of repetition, in which the having-been is alien, and is historical action which allows a new world emerge.

15 Cf. Heidegger, 2002, p. 26; Hw, pp. 33–4 [translation modified]: 'Only what moves can repose [*ruhen*]. The mode of repose [*Ruhe*] is determined by the mode of movement. In motion that is the mere change of place of a body, repose is, admittedly, only the limiting case of motion. When repose includes motion, there can be a repose which is an inner collection of motion. Such repose is, therefore, a state of extreme agitation – presupposing that the kind of motion in question requires such repose. The repose of the work that rests in itself is, however, of this sort. We will come, therefore, into the proximity of this repose if we can manage to grasp the movement of the happening in the work-being of the work as a unity.'

16 According to Heidegger, Heraclitus' understanding of 'polemos' is rooted in the relation of transformation and work. Cf. Heidegger, 2000, pp. 61–2; EiM, pp. 47–8.

17 Cf. Heidegger, 2002, pp. 40–1; Hw, pp. 52–3: 'To submit to this displacement [into the openness of entities opened up by the work] means: to transform all familiar relations to world and to earth and henceforth to restrain all usual doing and prizing, knowing and looking, in order to dwell with the truth that is happening in the work ... This allowing the work to be a work is what we call its preservation.'

He goes on: 'Preservation of the work means: standing within the openness of beings that happens in the work. This urgent standing-withinness [*Inständigkeit*] of preservation is, however, a knowing.'

18 Cf. ibid., p. 47; p. 61: 'equally poetic, though, in its own way, is the preservation of the work. For a work only actually is a work when we transport ourselves out of the habitual and into what is opened up by the work so as to bring out essence itself to take a stand within the truth of entities.'

19 Cf. ibid., pp. 2–3; Hw, p. 3 [translation modified]: 'works are objectively present as naturally as any other things …Works are shipped like coal from the Ruhr or logs from the Black Forest … [This is the] conception of the work with which the freight-handler or the museum charlady operates.'

20 *That* something is was named *existentia* by Thomists in the Middle Ages. In *Being and Time*, Heidegger seeks to show that the 'that it is' has constantly been interpreted as objective presence (cf. BT, p. 39; H, p. 42). The 'that is' of Dasein, however, is its thrownness and because the being of Dasein is not genuinely objectively present, but rather a movement (of projection and thrownness), that-being must be reinterpreted. When Heidegger now talks of 'that-being' in the Artwork Essays, he does so with respect to an entity which – in the terms of *Being and Time* – would be a non-Dasein like entity. This shows that Dasein – at least human Dasein – is no longer at the centre of Heidegger's concerns.

21 Cf. Heidegger, 2002, p. 39; Hw, p. 51: 'To be sure, 'that' it is made also belongs to every piece of equipment that is available for, and in, use. This 'that', however, is not salient in equipment; it disappears into usefulness.'

22 It is not accidental that leading political figures in both dictatorships and democracy have attempted to leave their mark in the arts, especially in architecture. This is true from Pericles and Augustus to Washington to Mussolini and Hitler, and more recently Françoise Mitterrand, all of whom have attempted to represent their political actions and aims in architecture.

23 Cf. Heidegger, 1998, pp. 127–8; Wm, pp. 163–4. Heidegger does not state directly that projection should be understood as a projection of a work, but projection is understood as the grounding of being in the sense of establishing (*Stiftung*).

24 BT, p. 347; H, p. 379: 'history is the specific occurrence of existing Dasein happening in time, in such a way that the occurrence in being-with-on-another is "past" and at the same time "handed down" and still having its effect is taken to be history in the sense emphasized.'

25 Dreyfus, 1984, p. 27, puts it well: '*Being and Time* has no place for the withdrawal and resistance of the earth'. Cf. also on this Harries, 1978, p. 320n. 33.

26 It is of course the case that Heidegger says expressly that the having-been can never be strange. This is so, however, only with respect to the project of fundamental ontology of *Being and Time*. Here we are dealing with the

having-beenness of the work, which remains to a certain extent strange to Dasein because the work is not founded in the making and preserving of Dasein (i.e. in its action), but rather the other way around: Dasein's action is founded in the work.

27 Cf. Heidegger, 1989a, p. 17: 'All art is in essence poetry, i.e. the opening up of that open in which everything is otherwise that it normally is.' Cf. also Heidegger, 2002, p. 46; Hw, p. 58. The basis of this strangeness lies in the movement of entities towards being, from the work as entity to the world, which is opened up in it.

28 The concept of 'metamorphosis' leads to a misunderstanding because the word itself suggests a form/matter schema from which Heidegger wishes to distance himself. To avoid such a consequence, I am emphasizing the temporal structure of metamorphosis. On metamorphosis in antiquity, see Forbes, 1992.

29 Cf. Heidegger, 2002, p. 42; Hw, p. 55 [translation modified]: 'In the earth … as the essentially self-secluding, the openness of the open encounters the highest form of resistance.'

30 Cf. Heidegger, 1989, p. 175: 'in one way or another we must take responsibility [*verantworten*] for that being to which we are delivered over [*überantwortet*]'

31 Again it is informative to look at the beginning of the 'Letter on "Humanism"', where this claim or calling to account is understood in terms of the ontological difference and through excluding causality: 'all effecting [*Wirken*] is directed towards entities. Thinking, in contrast, lets itself be claimed by being so that it can say the truth of being.' (Heidegger, 1998, p. 239; Hw, p. 311 [translation modified])

32 Cf. Heidegger, 1989, p. 192: 'The original belonging together is the ground for fidelity to be-ing [*Treue zum Seyn*]. Fidelity to be-ing is the precondition for all self-revealing relations to particular entities [*sich entfaltende, so und so seiende Verhalten*]. On the other hand, whoever departs the place easily, shows that he has no origin and is also only something objectively present.'

33 Cf. *Critique of Pure Reason*, B 116–7/A 84–5, where Kant introduces the deduction as pursuing the question of 'right' [*Recht*] with respect to concepts. Cf. on this theme Heidegger, 1990, pp. 57–8; KPM, pp. 85–6.

34 Cf. Heidegger, 1989, p. 202. Cf. Hölderlin, The Rhine, lines 29–30: 'for as he writhed without light in his fetters, terrible was the demi-god's raving [*da lichtlos er / In den Fesseln sich wälzte / Das Rasen des Halbgotts*]' (Hölderlin, 1986, p. 161)

35 *Überhören* means literally to 'overhear', where the prefix 'over' functions similarly to 'overlook' in English, meaning a failure to hear.

36 Michel Haar makes note of this difference, and as a consequence singles Angst out as an *ahistorical* mood, in contrast to the other fundamental moods (such as wonder, horror, joy, boredom, reticence) which he considers as historical. (cf. Haar, 1988, p. 268) After the turning, Angst plays for Heidegger a trans-epochal role, according to Haar. Angst is not a mood characteristic of any particular epoch, but rather brings human beings to the point of transition in which they do not

know themselves. (ibid., pp. 280–1) I am not convinced, however, that Angst plays such a unique role, rather I take it to be characteristic of all fundamental moods that they bring human being to such a point of transition. It is notable that Haar takes no account of the mood of sorrow, which Heidegger continually stresses to have the capacity to disclose transitions. Furthermore, access to transitions is not a way beyond history, but rather to its core. The possibility of transition is not beyond history as Haar claims (ibid., p. 277), bur rather constitutes the possibility of history considered kairologically.

37 Cf. 1989, p. 201: The poets 'harkening holds fast to the awfulness of the chained origin. This steadfast harkening is suffering [*Leiden*]. Suffering is, however, the being of the demi-god.'

38 Cf. Heidegger, 1989, pp. 217–18: 'Seen with respect to essence language in itself is the original poetry, and the poetic in it in the strict sense – that which we specifically call "poetry" – is the primordial language of a people, which proliferates as prose and in this proliferation is levelled off and degenerates'. Pol Vandevelde (1992) fails to recognize this task of poetry to bring the unique to language. His thesis in that article is that the primordial poetry [*Urdichtung*] in the first Hölderlin lecture course corresponds to discourse (*Rede*) in *Being and Time*, and that poetry as work is a poor imitation of this primordial poetry. Vandevelde understands the difference between discourse and language as the difference between silence and its contamination through innerworldly things. The difference, here, though is not the same (although it might lead as a consequence to the difference discussed by Vandevelde). Discourse in *Being and Time* concerns the unique, and Heidegger understands poetry in the first Hölderlin lecture course in exactly the same way . The setting in work of truth is not merely a setting in work which serves thinking in understanding, as Vandevelle suggests (ibid., p. 29), but rather a unique occurrence.

39 Cf. Heidegger, 1989, p. 23, where Heidegger contrast the textures of sound and vibration of poetic words and saying to the abstract truths of prosaic philosophical speaking.

40 Cf. also 1989, p. 70 'Our be-ing [*Seyn*] happens as conversation [*Gespräch*], in the occurrence of which the gods address us, place us under their claim, *bring us to language*.' [emphasis in original]

41 Cf. Heidegger, 1989, p. 95: 'Having to forego the old gods, the bearing of this forgoing, is the *preserving* of their divinity.' The point for Heidegger is to find the appropriate mode of acting, that is responsible action as a response to the flown gods. He returns to this theme in a number of places in the first lecture course on Hölderlin, e.g. ibid., pp. 84, 97, 98

42 Heidegger is not totally unambiguous on this point. On the one hand, he speaks of an unfulfilled godliness and the possibility of a 'new encounter with the gods'

(ibid., p. 97). On the other hand, he stresses the fundamental mood of sorrow of poetry which discloses precisely the not being able to transcend the conflict between proximity and distance of the gods. Divinity is a promise, which cannot be because it opens up a future which promises more than any possible present could contain.

43 Cf. Sitter, 1970, for what remains a compelling response to the decisionistic interpretation of Heidegger.

Chapter 5

1 Cf. Beistegui, 1998, p. 31: 'Is this not Heidegger's mistake, then: to have mistaken National Socialism for a revolution in the most genuine sense, to have misjudged it to the point of seeing it as an authentic relation to the power of the origin?' Beistegui's book is by far, in my view, the most clear sighted and genuine engagement with Heidegger's political engagement, because it more than any other sees the depth of the philosophical issues at play.

2 It is for this reason that Thomas Paine, in his response to Edmund Burke's critique of the French Revolution, suggested that the latter should really be understood as a 'counter-revolution' (cf. Koselleck, 1984, p. 373). It is in the midst of the French Revolution, perhaps with the Jacobin supremacy, that the meaning of the term 'revolution' changes to mean the arising of a new order, one which did not claim legitimacy as the restoration of an old order. It is undoubtedly the case that Heidegger displays little sympathy for the French Revolution as an event. In his interpretations of Hölderlin he does not, to my knowledge, give any weight to the influence the Revolution had on him. Furthermore, when at the beginning of his Schelling lecture course he speaks about the historical situation in 1809 he does so from an exclusively 'nationalistic' standpoint. Here, however, my concern is exclusively with the experience of revolution and how the transformation of the meaning of revolution – a transformation which we can trace to the French Revolution – moulds that experience.

3 Faye takes a different approach and denies that biologism was the sole or even primary source of Nazi racism. However, what that points to is the difficulty in pinning down any ideas which would form the basis of a Nazi ideology, something which makes it difficult to either excuse or convict someone of *ideological* sympathies with the Nazis. If the question is not ideological, then what is it? It seems to me the question at issue is the lack of philosophical obstacles to Nazism as a political practice, due not to Heidegger's perfidy, but to the exhaustion of philosophy as such. Cf. Faye, 2009, pp. 96–9.

4 If I may be allowed to quote from a different context and a different people: 'In

the long history of the world, only a few generations have been granted the role of defending freedom in its hour of maximum danger. I do not shrink from this responsibility – I welcome it. I do not believe that any of us would exchange places with any other people.' (http://avalon.law.yale.edu/20th_century/kennedy.asp [accessed 4th January, 2012]) These lines from John F. Kennedy's inaugural address reflect a similar understanding of mission of a people at a particular and kairological historical moment. This may seem an inappropriate comparison, as Faye in particular argues that Heidegger is putting philosophy at the disposal of the interests of a people and as such betraying it. That is a different argument, one to which I will return in section three; for now my point is simply that such claims regarding the mission of a people are not necessarily fascistic.

5 Heidegger himself approaches this question just a couple of pages after the line just quoted from the Artwork essay: 'to what extent is an impulse to something like a work contained in the essence of truth? What is the essence of truth, that it can be set into the work – even, *under certain conditions*, must be set into the work – in order to have its being as truth.' (Heidegger, 1987, p. 33; Hw, p. 41 [my emphasis])

6 Cf. Heidegger, 1987, p. 55; Hw, p. 71: 'Reflection on what *art* may be is completely and decisively directed solely toward the question of *being*.' In a contemporaneous text he states: 'In order to understand what the work of art and poetry as such are, the philosopher must first cease to think of the problem of art in aesthetic terms.' (Heidegger, 2002b, p. 47; GA 34, p. 64)

7 *Geschick* is mostly understood – and with good reason – as a gift or a sending. William Richardson (Richardson, 1974, p. 435) for example translates it with the neologism 'mittence'. Referring back to our discussion in Chapter 1, however, the ordering role of '*Geschick*' is emphasized here. In this sense, the destiny of being can be understood as the epochal bringing into order of entities.

8 On this question I find Stanley Rosen particularly insightful. Cf. his 'The Quarrel between Philosophy and Poetry' in Rosen, 1988, pp. 1–26.

9 The poet lies, according to Plato, but so does the philosopher in Plato's 'ideal' state. The latter's motivation may be different and he may be guided by his view of the ideas. But in the cave, where political action occurs, the difference between the action of the philosopher and that of the poet remains unclear.

10 Towards the end of the *Phenomenology of Spirit*, there is even the suggestion that the movement of spirit is cyclical such that at the end of history all must begin again: absorbed in itself [spirit] is sunk in the night of its self-consciousness; but in the night its vanished outer existence is preserved, and this transformed existence ... is the new existence, a new world and a new shape of spirit. In the immediacy of this new existence the spirit has to start afresh to bring itself to maturity'. (Hegel, 1977, p. 492) Cf. Gillespie, 1984, p. 114.

11 Joachim Heinrich Campe's recording of his experiences during the first months

of the French Revolution are revealing in this respect: 'one sees how the lowest citizen and the decorated man both appear as men and not in their respective official roles and walk along as fully equal with one another, without betraying any sign of impertinence on the part of one or offensive pride on the part of the other.' (Campe, 2011, p. 135 [my emphasis]). The letter containing these words is dated 9 August 1789.

12 Cf. Schwan, 1989, p. 87. Schwan interprets this 'at once' purely chronologically, as I see it. It is truer to Heidegger's sense to interpret it kairologically.

Bibliography

Primary texts (Heidegger)

Heidegger, M. (1977a), *The Question concerning Technology and Other Essays*. Trans. W. Lovitt. London Harper and Row; Heidegger (2011): *Die Technik und die Kehre*. Stuttgart: Klett-Cotta.

—(1977b), *Basic Writings*. D. Farrell Krell, (ed.) London: Harper and Row.

—(1982), *On the Way to Language*. Trans. P. Hertz. London: Harper & Row; Heidegger (2007) *Unterwegs zur Sprache*. Stuttgart: Klett-Cotta.

—(1984), *The Metaphysical Foundations of Logic*. Trans. M. Heim. Bloomington: Indiana University Press; Heidegger (1978): *Die metaphysiche Anfangsgründe der Logic. Gesamtausgabe 26*, K. Held. (ed.) Frankfurt a.M.: Klostermann.

—(1985a), *History of the Concept of Time*. Trans. T. Kisiel. Bloomington: Indiana University Press; Heidegger (1988): *Prolegomena zur Geschichte des Zeitbegriffs. Gesamtausgabe 20*, P. Jaeger. (ed.) Frankfurt a.M: Klosermann.

—(1985b), 'The Self-assertion of the German University. Address delivered on the Solemn Assumption of the Rectorate of the University of Freiburg. Rectorate 1933–4: Facts and Thoughts'. Trans. K. Harries in T*he Review of Metaphysics*, 38 (3), pp. 467–502; Heidegger (1983): *Das Rektorat 1933/34. Tatsachen und Gedanken*. Frankfurt a.M.: Klostermann.

—(1987), *Introduction to Metaphysics*. Trans. R. Manheim. New Haven: Yale University Press; Heidegger (1987): *Einführung in die Metaphysik*. Tübingen: Niemeyer.

—(1988), *The Basic Problems of Phenomenology*. Bloomington: Indiana University Press; Heidegger (1989): *Die Grundprobleme der Phänomenologie. Gesamtausgabe 24* (second edition), F.-W. von Hermann (ed.). Frankfurt a.M: Klosermann.

—(1989a), *Hölderlins Hymnen 'Germanien' und 'Der Rhein'. Gesamausgabe 39*, S. Ziegler, (ed.). Frankfurt a. M: Klosermann.

—(1989b), 'Von Ursprung des Kunstwerkes: Erste Ausarbeitung' in *Heidegger Studies*, 8, pp. 5–22.

—(1990), *Kant and the Problem of Metaphysics* (fourth edition). Trans. R. Taft. Bloomington: Indiana University Press; Heidegger (1991): *Heidegger und das Problem der Metaphysik*. Frankfurt a.M.: Klostermann.

—(1991a), *Logica: Leciones de M. Heidegger (semester verano 1934) en el legado de Helene Weiss*. Bilingual edition, V. Farias (ed.). Barcelona: Anthropos.

—(1991b), 'Unbenutzte Vorarbeiten zur Vorlesung vom Wintersemester 1929/30 'Die Grundbegriffe der Metaphysik. Welt – Endlichkeit – Einsamkeit' in *Heidegger Studies*, 7.

—(1994), *Basic Questions of Philosophy. Selected 'Problems' of 'Logic'* Trans. J. Stambaugh. Bloomington: Indiana University Press; Heidegger (1992) *Grundfragen der Philosophie. Ausgewählte „Probleme" der „Logik". Gesamtausgabe 45* (second edition), F.-W. von Herrmann (ed.). Frankfurt a.M.: Klostermann.

—(1995), *The Fundamental Concepts of Metaphysics. World, Finitude, Solitude.* Trans. W. McNeill and N. Walker. Bloomington: Indiana University Press; Heidegger (1992): *Die Grundbegriffe der Metaphysik. Welt – Endlichkeit – Einsamkeit. Gesamtausgabe 29/30*, F.-W. von Herrmann (ed.). Frankfurt a.M.: Klostermann.

—(1996), *Erläuterungen zu Hölderlins Dichtung* (6th edition). Frankfurt a.M.: Klostermann.

—(1998), *Pathmarks.* Trans. by W. McNeill, Cambridge: Cambridge University Press; Heidegger (1978): *Wegmarken* (second edition). Frankfurt a.M.: Klostermann.

—(1999), *Contributions to Philosophy: From Enowning.* Trans. P. Emad and K. Maly. Bloomington: Indiana University Press; Heidegger (1989): *Beiträge zur Philosophie. Von Ereignis. Gesamtausgabe 65.* Frankfurt a.M.: Klostrermann.

—(2000), *Towards a Definition of Philosophy.* Trans. T. Sadler. New Brunswick, Athlone; Heidegger (1999): *Zur Bestimmung der Philosophie. Gesamtausgabe 56/57* (second edition), B. Heimbüchel (ed.). Frankfurt a.M.: Klostermann.

—(2001), *Zollikon Seminars. protocols, conversations, letters.* Trans. F. Mayr and R. Askay. Evanston: Northwestern University Press; Heidegger (1994): *Zollikoner Seminare Protokolle, Zwiegespräche, Briefe*, M. Boss (ed.). Frankfurt a.M.: Klostermann.

—(2002a), *Off the Beaten Track.* Trans. J. Young. Cambridge: Cambridge University Press; Heidegger (1980): *Holzwege.* Frankfurt a.M.: Klostermann.

—(2002b), *The Essence of Human Freedom: An Introduction to Philosophy.* Trans. T. Sadler. London: Continuum Books; Heidegger (1982): *Vom Wesen der menschlichen Freiheit. Gesamtausgabe 31*, H. Tietjen (ed.). Frankfurt a.M.: Klostermann.

—(2002c), *The Essence of Truth: On Plato's Parable of the Cave and the Theaetus.* Trans. T. Sadler. London: Continuum Books; Heidegger (1988): *Vom Wesen der Wahrheit. Gesamtausgabe 34.* H. Mörchen, (ed.). Frankfurt a.M.: Klostermann.

—(2003), *Plato's Sophist.* Trans. R. Rojcewicz and A. Schuwer. Bloomington: Indiana University Press; Heidegger (1992): *Platons Sophistes. Gesamtausgabe 19*, I. Schüßler (ed.). Frankfurt a.M.: Klostermann.

—(2004), *The Phenomenology of Religious Life.* Trans. M. Fritsch and J. Gosetti-Ferencei. Bloomington: Indiana University Press; Heidegger (1995): *Phänomenologie des religiösen Lebens. Gesamtausgabe 60*, M. Jung, T. Regehly and C. Strube (eds). Frankfurt a.M.: Klostermann.

—(2006), *Being and Time: a Translation of Sein und Zeit.* Trans. J. Stambaugh. Albany: SUNY Press; Heidegger (1986) *Sein und Zeit* (seventh edition). Tübingen: Niemeyer.

—(2007a), 'The Concept of Time in the Science of History', in T. Kisiel and T. Sheehen: *Becoming Heidegger: on the trail of his early occasional writings, 1910-1927.*

Evanston: Northwestern University Press, pp. 60–72; Heidegger (1978): 'Der Zeitbegriff in der Geisteswissenschaften', in Heidegger (1978): *Frühe Schriften (1912-1916). Gesamtausgabe vol. 1*, F.-W. von Hermann (ed.). Frankfurt a.M.: Klostermann, pp. 415–33

—(2007b), 'The Concept of Time', in T. Kisiel and T. Sheehen: *Becoming Heidegger: on the trail of his early occasional writings, 1910–1927*. Evanston: Northwestern University Press, pp. 196–213; Heidegger (2004): 'Der Begriff der Zeit', in *Der Begriff der Zeit. Gesamtausgabe 64*, F.W. von Hermann (ed.). Frankfurt a.M.: Klostermann, pp. 105–25.

—(2007c), 'Wilhelm Dilthey's Research and the Current Struggle for a Historical Worldview', in T. Kisiel and T. Sheehen: *Becoming Heidegger: on the trail of his early occasional writings, 1910–1927*. Evanston: Northwestern University Press, pp. 238–74; Heidegger (1992/3): 'Kasseler Vorträge', Bröcker and Rodi (eds). *Dilthey Jahrbuch*, 8, pp. 143–80.

—(2009), *Logic as the Question concerning the Essence of Language*. Trans. W. Gregory and Y. Unna. Albany: SUNY Press; Heidegger (1998): *Logik als Frage nach des Wesens der Sprache. Gesamtausgabe 38*, G. Seubold (ed.). Frankfurt a.M.: Klostermann.

—(2010), *Logic: the question concerning truth*. Trans. T. Sheehen. Bloomington: Indiana University Press; Heidegger (1995): *Logic: die Frage nach der Wahrheit. Gesamtausgabe 21*, W. Biemel (ed.). Frankfurt a.M.: Klostermann.

Secondary works on Heidegger

Barash, J. (2003), *Martin Heidegger and the Problem of Historical Meaning*. Fordham: Fordham University Press.

Bernasconi, R. (1989), 'Heidegger's Destruction of Phronesis', *The Southern Journal of Philosophy*, 28 (Supplement), pp. 127–47.

—(1994), 'The Fate of the Distinction between Praxis and Poiesis' in *Heidegger in Question: The Art of Existing*. New York: Prometheus Books, pp. 2–24.

Bradner, R. (1994), *Heideggers Begriff der Geschichte und das neuzeitliche Geschichtsdenken*. Vienna: Passagen Verlag.

Crowell, S. (2001), *Husserl, Heidegger and the Space of Meaning. Paths toward Transcendental Phenomenology*. Evanston: Northwestern University Press.

—(2007), 'Conscience and Reason: Heidegger and Grounds of Intentionality', in S. Crowell and J. Malpas: *Transcendental Heidegger*. Stanford: Stanford University Press, pp. 43–62.

Dahlstrom, D. (2001), *Heidegger's Concept of Truth*. Cambridge: Cambridge University Press.

Dastur, F. (1998), *Heidegger and the Question of Time*. Trans. F. Raffoul. New York: Prometheus Books.

Derrida, J. (1989), *Of Spirit: Heidegger and the Question*. Trans. G. Bennington and R. Bowlby. Chicago: University of Chicago Press.

Dreyfus, H. (1984), 'Between Techne and Technology: The Ambiguous Place of Equipment in *Being and Time*', in Tulane Studies in Philosophy, 32, pp. 23–35.

Figal, G. (1988), *Martin Heidegger: Phänomenologie der Freiheit*. Frankfurt a.M.: Athenaum.

Fynsk, C. (1993), *Heidegger: Thought and Historicity*. Cornell: Cornell University Press.

Gethmann, G. F. (1989), 'Heideggers Konzeption des Handelns in Sein und Zeit', in A. Gethmann-Siefert and O. Pöggeler: *Heidegger und die praktische Philosophie*. Frankfurt a.M.: Suhrkamp, pp. 140–76.

Gillespie, M. (1984), *Hegel, Heidegger and the Ground of History*. Chicago: University of Chicago Press.

Gosetti-Ferencei, J. (2004), *Heidegger, Hölderlin and the Subject of Poetic Language: Towards a New Poetics of Dasein* (third edition). New York: Fordham University Press.

Grondin, J. (1987), *La tournant dans la pensée de Martin Heidegger*. Paris: Presses Universitaires de France.

Haar, M. (1988), 'Stimmung et Pensée' in F. Volpi (ed.): *Heidegger et l'Idée de la Phénomenologie*. Dordrecht: Springer.

Habermas, J. (1993), 'On the Publication of the Lectures of 1935' in R. Wolin (ed.), *The Heidegger Controversy: A Critical Reader*. Cambridge, MA: MIT Press, pp. 186–97.

Harman, G. (2002), *Tool-Being Heidegger and the Metaphysics of Objects*. Chicago: Open Court.

—(2007), *Heidegger Explained: From Phenomenon to Thing*. Chicago: Open Court

Harries, K. (1978), 'Heidegger as a Political Thinker' in M. Murray (ed.): *Heidegger and Modern Philosophy*. New Haven: Yale University Press, pp. 304–28.

Held, K. (1991), 'Grundstimmung und Zeitkritik bei Heidegger' in D. Papenfuss and O. Pöggeler (eds): *Zur philosophischen Aktualität Heideggers. Bd. 1 Philosophie und Politik*. Frankfurt a.M.: Klostermann, pp. 31–56.

Hodge, J (1995), *Heidegger and Ethics*. London: Routledge.

Hoy, D. (1978), 'History, Historicity and Historiography in *Being and Time*' in M. Murray: *Heidegger and Modern Philosophy*. New Haven: Yale University Press, pp. 329–53.

Hüni, H. (1985), 'Phänomenologie des Gewissens in Zusammenhang von Sein und Zeit' in *Perspektiven der Philosophie*, 2, pp. 31–46.

—(1993), Die Verhaltenheit der Natur nach Heidegger' in K. Held and J. Hennigfeld (eds): *Kategorien der Existenz. Festschrift für Wolfgang Janke*. Würzburg: Könighshausen und Neumann, pp. 297–304.

Jasspers, K. (1993), Letter to the Denazification Committee', in R. Wolin (ed.), *The Heidegger Controversy: A Critical Reader*. Cambridge, MA: MIT Press, pp. 144–51.

Kisiel, T. (1992), 'Heidegger's Apology: Biography as Philosophy and Ideology', in

T. Rockmore (ed.): *The Heidegger Case: On Philosophy and Politics*. Philadelphia: Temple University Press, pp. 11–51.

—(1993), *The Genesis of Heidegger's Being and Time*. Berkeley: University of California Press.

Lacoue-Labarthe, P. (1990), *Heidegger, Art, and Politics: The Fiction of the Political*. London: Basil Blackwell.

Löwith, K. (1993), 'My Last Meeting with Heidegger in Rome, 1936', in R. Wolin (ed.), *The Heidegger Controversy: A Critical Reader*. Cambridge, MA: MIT Press, pp. 140–3.

McNeill, W. (1999), *Glance of the Eye: Heidegger, Aristotle and the Ends of Theory*. Albany: SUNY Press.

—(2006), *The Time of Life: Heidegger and Ěthos*. Albany: SUNY Press.

Müller-Lauter, W. (1960), *Möglichkeit und Wirklichkeit bei Martin Heidegger*. Berlin: de Gruyter.

Ó Murchadha, F. (1998), 'Future or Future Past: Temporality between Praxis and Poiesis in Heidegger's *Being and Time*', *Philosophy Today*, 42 (3), pp. 262–9.

—(2012), "The Political and Ethical Significance of Waiting: Heidegger and the Legacy of Thinking" in F. Halsall, J. Jansen and S. Murphy (eds): *Critical Communities and Aesthetic Practices: Dialogues with Tony O' Connor on Society, Art and Friendship*. Dordrecht:Springer Press, pp. 139–50.

Pöggeler, O. (1985), 'Den Führer führen? Heidegger und kein Ende', in *Philosophische Rundschau*, 32, pp. 26–67.

—(1989), *Martin Heidegger's Path of Thinking*. Trans. by D. Magurshak and S. Barber. Atlantic Highlands: Humanities Press.

—(1990), 'Philosophie und Nationalsozialismus. Am Beispiel Heideggers' in *Rheinisch-Westfälisch Akademie der Wissenschaft Vorträge*. Opladen: Westdeutscher Verlag.

—(1994), 'Destruction and Moment' in T. Kisiel and J. van Buren: *Reading Heidegger from the Start: essays in his earliest thought*. Albany: SUNY Press.

Prauss, G. (1999), *Doing and Knowing in Heidegger's Being and Time*. Trans. G. Steiner and J. Turner, London: Humanities Books.

Richardson, W. (1974), *Heidegger: Through Phenomenology to Thought*. Dordrecht: Martinus Nijhoff.

Ruin, H. (1994), *Enigmatic Origins. tracing the theme of historicity through Heidegger's works*. Stockholm: Almqvist & Wiksell.

Schürmann, R. (1990), *Heidegger on Being and Acting: From Principles to Anarchy*. Bloomington: Indiana University Press.

Schwan, A. (1989), *Politische Philosophie in Denken Heideggers*. Opladen: Westdeutscher Verlag.

Sheehen, T. (1988), 'Heidegger and the Nazis' in *The New York Review of Books*, June 16, pp. 38-47.

Sluga, H. (1993), *Heidegger's Crisis: Philosophy and Politics in Nazi Germany*. Cambridge: Cambridge University Press.

Taminiaux, J. (1991), *Heidegger and the Project of Fundamental Ontology*. Trans. M. Gendre. New York: SUNY Press.
Tugendhat, E. (1970), *Der Wahrheitsbegriff bei Husserl and Heidegger*. Berlin: de Gruyter.
Volpi, F. (1984), Heidegger in Marburg: die Auseinandersetzung mit Aristoteles' in *Philosophische Literaturanzeige*, 37 (2), pp. 172–88.
—(1988), 'Dasein comme Praxis' in Volpi et al. (ed.): *Heidegger et le idee de la phénomenologie*. Dordrecht: Springer, pp. 1–42.
Young, J. (1997), *Heidegger, Philosophy, Nazism*. Cambridge: Cambridge University Press.
Zangeheh, H. (2011), 'Phenomenological Problems for the Kairological Reading of Augenblick in *Being and Time*'. *International Journal of Philosophical Studies* 19 (4), pp. 539–61.
Ziegler, S. (1991), *Heidegger, Hölderlin und die 'Αλήθεια. Martin Heideggers Geschichtsdenken in seinen Vorlesungen 1934/5 bis 1944*. Berlin: Dunker und Humblot.
Zimmermann, M. (1990), *Heidegger's Confrontation with Modernity: Technology, Politics, Art*. Bloomington: Indiana University Press.

Other works cited

Arendt, H. (1968), *Between Past and Future*. London: Penguin.
—(1973), *On Revolution*. London: Penguin.
—(1978), *The Life of the Mind: On Willing*. New York: Harcourt Brace.
—(1998), *The Human Condition*. Chicago: University of Chicago Press.
Aristotle (1984), *The Complete Works of Aristotle: the revised Oxford translation*. Ed. J. Barnes. Oxford: Oxford University Press.
Aubenque, P. (1976), *Le Prudence chez Aristoteles*. Paris: PUF.
Augustine (1998), *The City of God*. Trans. R. Dyson. Cambridge: Cambridge University Press.
—(2006), *The Confessions* (Second Hackett edition). Trans. F.J. Sheed. London: Hackett.
Beierwaltes, W. (1966/7), '*Exaiphnes* oder die Paradoxie des Augenblicks' in *Philosophisches Jahrbuch*, 74, pp. 271–83.
Bergson, H. (1963), *The Two Sources of Morality and Religion*. South Bend: University of Notre Dame Press.
—(1968), *The Creative Mind*. Trans. M. Andison. New York: Greenwood Press.
Blumenberg, H. (1986), *Lebenszeit und Weltzeit*. Frankfurt a.M.: Suhrkamp.
Bohrer, K. H. (1981), *Plötzlichkeit. Zum Augenblick des ästhetischen Scheins*. Frankfurt a.M.: Suhrkamp.

Bubner, R. (1984), *Gesichtsprozesse und Handlungsnormen. Untersuchungen zur praktischen Philosophie*. Frankfurt a.M.: Suhrkamp.

Burkhardt, J. (1952), *Griechische Kulturgeschichte Bd 1. Der Staat und Religion*. Stuttgart. Deutsche Verlags-Anstalt.

Campe, J. (1961), *Briefe aus Paris während der Französische Revolution*, H. König (ed.). Berlin: Nabu Press.

Carr, D. (1986), *Time, Narrative and History*. Bloomington: Indiana University Press.

Davidson, D. (1980), *Essays on Actions and Events*. Oxford: Oxford University Press.

Deleuze, G. (2005), *The Logic of Sense*. Trans. C. Boundas. London: Continuum.

Derrida, J. (2008), *The Animal that Therefore I am* (third edition) Trans. D. Wills. New York: Fordham University Press.

Descartes, R. (1984), *The Philosophical Writings of Descartes vol. 2*, J. Cottingham, R. Stoffhoff and D. Murdoch (eds). Cambridge: Cambridge University Press.

Dilthey, W. (1957), 'Über das Studium der Geschichte des Wissenschaften von Menschen, der Gesellschaft und dem Staat' in *Gesammelte Schriften. Vol. 5* Stuttgart: Vandenhoeck und Ruprecht.

Dilthey, W. and Yorck, P. (1923), *Briefwechsel zwischen Wilhelm Dilthey und dem Grafen Paul Yorck von Wartenburg (1877–1897)*. Halle: Niemeyer.

Fink, E. (1960), *Spiel als Weltsymbol*. Stuttgart: Kohlhammer.

Flasch, K. (1993), *Was ist Zeit?* Frankfurt .a.M.: Klostermann.

Forbes, P. (1992), *Metamorphosis in Greek Myths*. Oxford: Oxford University Press.

Furet, F. (1981), *Interpreting the French Revolution*. Trans. E. Forster. Cambridge: Cambridge University Press.

Gadamer, H.-G. (1976), 'The Universality of the Hermeneutical Problem' in *Philosophical Hermeneutics*. Berkeley: University of California Press, pp. 3–17.

—(1993), 'Die Kontinuität der Geschichte und der Augenblick der Existenz', *Gesammelte Werke, Band 2*. Tübingen: Mohr Siebeck.

—(1999), 'Friendship and Self-Knowledge: Reflections on the Role of Friendship in Greek Ethics', in *Hermeneutics, Religion, and Ethics*. Trans. J. Weinsheimer New Haven, Co.: Yale University Press, 1999.

Hamilton, E. and Cairns, H. (1961), *The Collected Dialogues of Plato*. Princeton: Princeton University Press.

Hartley, L. P. (2002), *The Go Between*. New York: New York Review of Books.

Hegel, G. W. F. (1970), *Vorlesungen über Ästhetik. Werke 7*. Frankfurt a.M.: Suhrkamp.

Held, K. (1991), 'Heimwelt, Fremdwelt, die eine Welt', in *Phänomenologische Forschungen*, 24/25, pp. 305–37.

—(1993), 'Eigentliche Existenz und politische Welt' in K. Held and J. Hennigfeld (eds): *Kategorien der Existenz. Festschrift für Wolfgang Janke*. Würzburg: Könighshausen und Neumann, pp. 395–412.

—(1996), 'Generative Zeiterfahrung' in *Edith Stein Jahrbuch*, 2, 265–82.

Hölderlin, F. (1986), *Selected Poems*. Trans. by M. Hamburger. London: Anvil Press.

Homer (1998), *The Iliad*. Trans. R. Fitzgerald. Oxford: Oxford University Press.

Husserl, E. (1964), *The Phenomenology of Internal Time Consciousness*. Trans. J. Brough. The Hague: Martinus Nijhoff.

Jaspers, K. (1989), *Psychologie der Weltanschauungen*. Berlin: Springer.

Kafka, F. (1937), *Novellen, Skizzen, Aphorismen aus dem Nachlass*. Frankfurt a.M.: Fischer.

Kant, I. (1998), *Critique of Pure Reason*. Trans. P. Guyer and A. Wood. Cambridge: Cambridge University Press.

Kearney, R. (1984), *Poetique du Possible*. Paris: Beuchesne.

Kierkegaard, S. (1962), *Philosophical Fragments*. Trans. D. Swenso. Princeton: Princeton University Press.

—(1980), *The Concept of Anxiety*. Trans. R. Thormte. New Haven: Princeton University Press.

—(1983), *Fear and Trembling/Repetition: Kierkegaard's Writings, Vol. 6*. Trans. E. Hong and H. Hong. Princeton: Princeton University Press.

Koselleck, R. (1984), 'Revolution' in O. Brunner, W. Conze and R. Koselleck: *Geschichtliche Grundbegriffe. Historisches Lexikon zur politisch-sozialisch Sprache in Deutsch, vol. 5*. Stuttgart: Klett-Cotta.

—(1985), *Futures Past*. Trans. K. Tribe. Cambridge, MA: MIT Press.

Lawlor, L. (2007), *This is not Sufficient. An Essay on Animality and Human Nature in Derrida*. New York: Columbia University Press.

Levi-Strauss, C. (1966), *The Savage Mind*. Chicago: University of Chicago Press.

Levinas, E. (1969), *Totality and Infinity*. Trans. A. Lingis. Pittsburgh: Duquesne University Press.

Link, C. (1984), 'Der Augenblick. Das Problem des platonischen Zeitverständnisses' in *Die Erfahrung der Zeit. Gedankenschrift für Georg Picht*. Stuttgart: Klett-Cota, pp. 51–84.

Löwith, K. (1986), *Mein Leben in Deutschland vor und nach 1933*. Stuttgart: Metzler.

Marx, K. (1977), 'The Eighteenth Brumaire of Louis Napoleon' in *Selected Works. Volume 1*. Moscow: Progress Publishers, pp. 398–487.

Meier, C. (1984), 'Revolution in der Antike' in O. Brunner, W. Conze and R. Koselleck: *Geschichtliche Grundbegriffe. Historisches Lexikon zur politisch-sozialisch Sprache in Deutsch, vol. 5*. Stuttgart: Klett-Cotta.

Nancy, J. L. (1991), *The Inoperative Community*. Trans. P. Connor, L. Garbus, M. Holland and S. Sawhney. Minneapolis: University of Minnesota Press.

Nietzsche, F. (1997), *Untimely Meditations*. Trans. R. J. Holingdale. Cambridge: Cambridge University Press.

—(2001), *The Gay Science*. Trans. J. Nauckhoff and A. del Caro. Cambridge: Cambridge University Press.

O'Byrne, A. (2010), *Natality and Finitude*. Bloomington: Indiana University Press.

Plato (1989), *The Collected Dialogues of Plato*, E. Hamilton (ed.). Princeton: Princeton University Press.

Renthe-Fink, C. (1964), *Geschichtlichkeit. Ihr terminologischer und begrifflicher Ursprung bein Hegel, Haym, Dilthey und Yorck*. Göttingen: Vandenhoeck und Ruprecht.

Ricoeur, P. (1984), *Time and Narrative vol. 1*. Trans. K. McLaughlin and D. Pellauer. Chicago: University of Chicago Press.
Ritter, J. (1977), *Historisches Wörterbuch der Philosophie vol. 4*. Basel: Schwabe.
Rosen, S. (1988), *The Quarrel Between Philosophy and Poetry*. London: Routledge.
Saner, M. (1976), 'Momento nasci. Vorbemerkung zu einer Philosophie der Geburt' in G.K. Kaltenbrunner (ed.): *Überleben und Ethik. Die Notwendigkeit bescheiden zu werden*. Freiburg i. Br.: Alber.
Schwan, A. (1963), 'Politik als "Werk der Wahrheit". Einheit und Differenz von Ethik und Politik bei Aristoteles' in P. Engelhardt (ed.): *Sein und Ethos*. Mainz: Matthias-Grünwald-Verlag, pp. 69–110.
Simmel, G. (1997), *Simmel on Culture: Selected Writings*, D. Frisby and M. Featherstone (eds). London: Sage Publications.
Sitter, B. (1970), 'Zur Möglichkeit dezisionistischer Auslegung von Heideggers ersten Schriften' in *Zeitschrift für philosophische Forschung*, 24, pp. 516–35.
Sommer, M. (1990), *Lebenswelt und Zeitbewußtsein*. Frankfurt a.M.: Suhrkamp.
Theunissen, M. (1991), *Die negative Theologie der Zeit*. Frankfurt a.M.: Suhrkamp.
Tillich, P. (1963), *Der Widerstreit von Raum und Zeit. Gesammelte Werke VI*. Stuttgart: Evangeliches Verlagswerk.
Tocqueville, A. de (1971), *The Ancient Régime and The French Revolution*. Trans. S. Gilbert. London: Fontana.
Trédé, M. (1984), 'Καιρός: problèmes d'étymologie' *Reve des Etudes Grecques*, XCVII, pp. xi–xvi.
Vuillemin, J. (1984), *Necessite ou Contingence: l'aporie de Diodore et les systems philosophiques*. Paris: Minuit.
Waldenfels, B. (1987), *Ordnung in Zwielicht*. Frankfurt a.M.: Suhrkamp.
—(1990), *Der Stachel des Fremden*. Frankfurt a.M.: Suhrkamp.
White, H. (1987), *The Content of the Form: Narrative Discourse and Historical Representation*. Baltimore: Johns Hopkins University Press.
Wohlfart, G. (1982), *Der Augenblick: Zeit und ästhetisiche Erfarhrung bei Kant, Hegel, Nietzsche und Heidegger mit einem Exkurs zu Proust*. Freiburg i. B: Alber.
Wörner, M. (2011), 'Die Zeit des Politischen in der Rhetorik des Aristoteles', in D. Pinter and U. Schubert (eds): *Wirtschaft - Gesellschaft - Natur*. Marburg: Metropolis Verlag.
Zakira, E. (1994), *La Causalité de Galilée à Kant*. Paris: Presse Universitaire de France.

Reference works

Drosdowski, G. (1963), *Duden Herkunftwörterbuch* Mannheim: Duden.
Grimm, J. and W. (1854ff), *Grimmisches Wörterbuch*. Leipzig.
New Jerusalem Bible: London, Darton, Longman & Todd, 1990.
http://avalon.law.yale.edu/20th_century/kennedy.asp (accessed 4th January 2012).

Index